D0595969

Mystifying Mathematical Puzzles

Mystifying Mathematical Puzzles

Golden Spheres, Squared Eggs and Other Brainteasers

Joseph S. Madachy

Illustrations by the Author

DOVER PUBLICATIONS, INC.
Mineola, New York

Bibliographical Note

This Dover edition, first published in 2018, is an unabridged and corrected republication of the work originally published by Charles Scribner's Sons, New York, in 1966 under the title *Mathematics on Vacation*. The work was previously reprinted by Dover in 1979 under the title *Madachy's Mathematical Recreations.*

Library of Congress Cataloging-in-Publication Data

Names: Madachy, Joseph S., author. | Madachy, Joseph S. Madachy's mathematical recreations.
Title: Mystifying mathematical puzzles : golden spheres, squared eggs and other brainteasers / Joseph S. Madachy ; illustrations by the author.
Other titles: Mathematics on vacation
Description: Dover edition. | Mineola, New York : Dover Publications, Inc., 2018. | An unabridged and corrected edition, originally published: New York : Charles Scribners Sons, 1966, under the title Mathematics on vacation. Previously republished by Dover Publishing in 1979, under the title Madachy's mathematical recreations.
Identifiers: LCCN 2017052959 | ISBN 9780486825076 | ISBN 0486825078
Subjects: LCSH: Mathematical recreations.
Classification: LCC QA95 .M27 2018 | DDC 793.74—dc23
LC record available at https://lccn.loc.gov/2017052959

Manufactured in the United States by LSC Communications
82507801 2018
www.doverpublications.com

ACKNOWLEDGEMENTS

Throughout this book I have given credit to those who contributed major portions of the material and results. I would like to thank the following for granting permission to include their contributions: W. H. Cozens, Douglas A. Engel, Donald E. Knuth, J. A. Lindon, William H. McGrail, Wade E. Philpott, William R. Ransom, Sidney H. Scott, and David F. Smith. In addition, I would like to thank Howard H. Bergerson, A. G. Bradbury, Steven R. Conrad, D. C. Cross, Clifford R. Dickinson, Alan Gold, Harvey Hahn, Jack Halliburton, R. H. Hide, J. A. H. Hunter, Jonathan Khuner, Sidney Kravitz, N. A. Longmore, Paul R. McClenon, Derrick Murdoch, Harry L. Nelson, Tom Rieder, W. A. Robb, Margaret M. Rohe, C. R. J. Singleton, and Anneliese Zimmerman for the many alphametics and puzzles that appear in Chapters 5 and 7.

Full responsibility for the accurate reproduction of all the contributed material in these pages is accepted by myself alone.

.: preface to the first edition

If you have ever solved a mathematical puzzle, played any game in which numbers were used, learned a number trick, or played ticktacktoe, you have indulged in a type of mathematical recreation. In general, these recreations share three characteristics: first, they are mathematical or logical; second, they are fun; third, they all appear to be quite useless.

The first two characteristics require no support, but some might ask why anyone should indulge in anything that is of no practical value. An indirect defense may be built up from the fact that the most staid mathematics and engineering journals publish some material that is strictly recreational in nature, giving the amusements a kind of endorsement.

Moreover, a very cursory examination will disclose a few interesting facts about the utility of recreational mathematics. Prime numbers, for example, have no practical value. It may be decades, if ever, before such a use for them will be found. However, the study of prime numbers and their properties has filled many a gap in the field of *number theory,* that mathematical discipline which studies the basic properties of all numbers. Magic squares have been involved with superstitious beliefs and proven an interesting source of amusement for centuries. They, too, offer practical rewards for the agricultural and nuclear scientist. The study of the patterns of certain types of magic square has shown how to reduce the number of experiments required to obtained growth and radiation data. The classic Moebius strip has been used for conveyor belts that last twice as long as conventional belts. Indeed, the B. F. Goodrich Co. has obtained a patent for this particular use. Many areas of recreational mathematics still seem to be useless, but who can say what next year or the next decade will bring?

This book attempts to give a sampling of both types of material. Much of its material is taken from the pages of *Recreational Mathematics Magazine,* which I founded, edited, and published from 1960 until its discontinuation in 1964. Moreover, many notes and comments from the periodical's readers as well as a number of original ideas have been incorporated here.

Thanks and direct acknowledgement to the many who contributed or helped will be found in the Acknowledgements and throughout the book. Thanks must also go to J. A. H. Hunter, Howard C. Saar, and Dmitri E. Thoro who, as co-editors of *Recreational Mathematics Magazine,* worked far beyond the call of duty. The author wishes to express his gratitude as well to the nearly ten thousand former subscribers whose enthusiasm and support helped make this book possible.

Kettering, Ohio J. S. M.

.: preface to the dover edition

This edition contains a little additional material, some corrections, and new statements to bring things up to date.

Many thanks go to many friends and readers who pointed out errors, informed me of recent developments, and suggested changes and modifications in the presentation of some of the problems and solutions.

Kettering, Ohio J. S. M.
June 1978

contents

Mystifying Mathematical Puzzles

1

geometric dissections

The volume will get off to a gentle start with a mathematical recreation that requires almost no work with numbers, and, as will be shown in other chapters, it is hardly alone in occupying such an apparently contradictory position. *Geometric dissections* involve the cutting up of any geometrical figure in some specified manner. For example, rectangles and squares can be dissected into any number of smaller, unequal squares or smaller, unequal rectangles. A similar dissection is that of a square or an *obtuse-angled triangle,* a triangle in which one angle is more than 90°, into a minimum number of acute-angled triangles, those whose angles are all less than 90°.

Until 1938, it was believed that the dissection of a square into smaller, unequal squares had no solution. In that year, R. L. Brooks, C. A. B. Smith, A. H. Stone, and W. T. Tutte, members of the Trinity Mathematical Society of Trinity College, England, succeeded in finding solutions. A typical solution to this problem is shown in Figure 20.

While this kind of diversion is quite interesting and affords the practitioner much amusement, this chapter will deal mainly with another type of geometric dissection: the conversion of one figure to another by the straightforward method of cutting one into a finite number of pieces and rearranging them to form the other. There is a certain aesthetic and mathematical satisfaction in such a procedure. The new figure, of course, has the same area as the original, and the sense of accomplishment is heightened when the transformation is performed by a dissection involving the least possible number of pieces.

It has been known for some time that any rectilinear plane figure, that is, any *polygon*, can be transformed into any other rectilinear plane figure of equal area by cutting it into some finite number of pieces. Any polygon, A, can be cut into triangles by drawing diagonals from one vertex to each of the other vertexes, and these triangles can be transformed into rectangles such that all have bases with the same length. These rectangles can be joined together to form a larger rectangle, A'. Any other polygon, B, with the same area as the first, but of a different shape, can also be divided into triangles by drawing diagonals from vertex to vertex. Again, these triangles can be transformed into rectangles having the same base as those formed from the triangles of polygon A, and another large rectangle, B', can be assembled by combining these smaller triangles. The two large rectangles A' and B' will be *congruent;* that is, one can be superimposed on the other and the two will coincide at all points. Now, A' can be subdivided into the same number of rectangles as there were triangles in polygon B. If these rectangles are made the same size as those that resulted from B, they can be transformed to triangles of the same size as those originally derived from it. Combining these triangles into B completes the transformation from A to B. Unfortunately, this proof offers no assistance whatever if it is desired to find the *minimum* number of pieces required for the transformation from A to B.

No calculations were needed for this transformation, but geometric dissections do sometimes require mathematical knowledge, which is usually used merely to establish that a given dissection of one figure does indeed produce the new one. For instance, when a rectangle-to-square transformation is made, calculations of lengths and areas should be made to assure that a true square results instead of a rectangle that is only very close to being a square. Eyesight is not to be trusted.

This field has a very long history. For example, a dissection in which a square is transformed into two smaller, unequal squares is used in an ancient proof of the *Pythagorean theorem*—a geometrical theorem stating that in the case of right-angled triangles, the square on the *hypotenuse,* or side opposite the right angle, is equal to the sum of the squares on the other

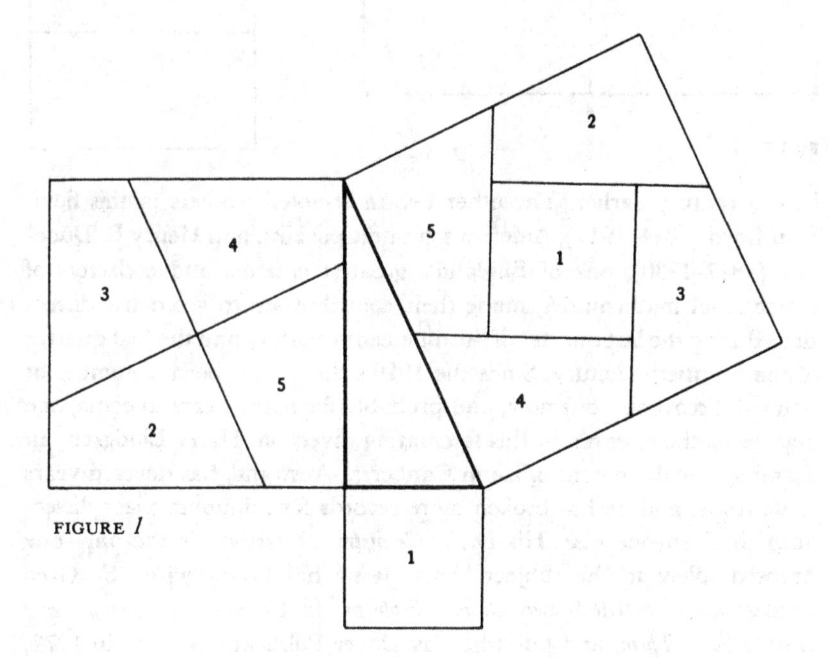

FIGURE *1*

2 sides—utilizing geometric dissection, as shown in Figure 1. The pieces forming the large square on the hypotenuse are used to form the 2 smaller squares on the other sides of the triangle.

Transformation of squares to regular *hexagons,* six-sided polygons, or *heptagons,* seven-sided polygons, were known at the beginning of the nineteenth century, whereas rectangle-to-square dissections were described by the French mathematician Jean Étienne Montucla (1725–1799) at

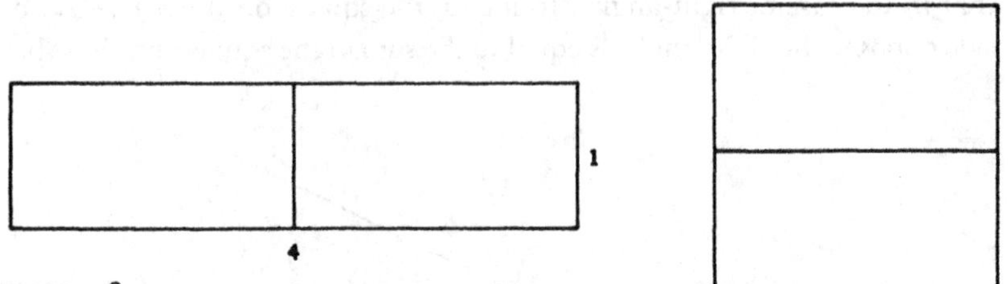

FIGURE 2

least a century earlier. The other two most-noted workers in this field, Sam Loyd (1844–1911), America's greatest puzzlist, and Henry E. Dudeney (1847–1930), one of England's greatest creators and collectors of recreational mathematics, made their contributions to geometric dissections during the last quarter of the nineteenth century and the first quarter of the twentieth century. Since the 1920's there have been a number of isolated discoveries, but now, and probably for many years to come, one man leads the research in this fascinating diversion. Harry Lindgren, an examiner for the patent office in Canberra, Australia, has devoted years to its study, and he has broken more records for minimum-piece dissections than anyone else. His book, *Geometric Dissections,* is the only one devoted solely to the subject. This classic has been revised by Greg Frederickson, retitled *Recreational Problems in Geometric Dissections and How to Solve Them*, and published by Dover Publications, Inc., in 1972.

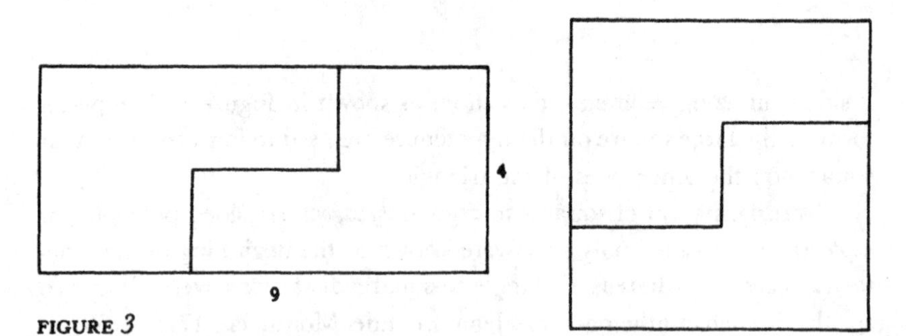

FIGURE 3

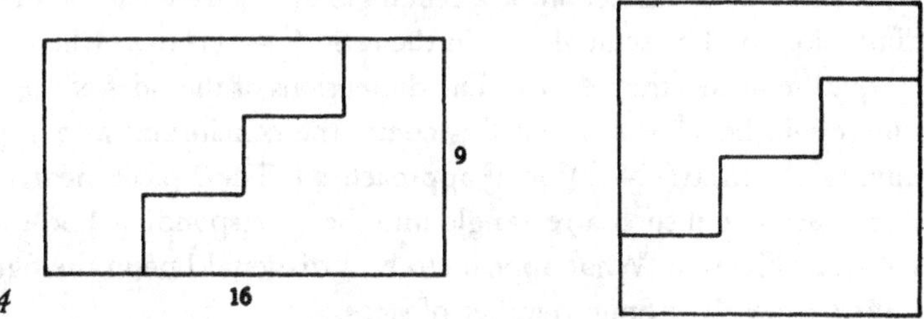

FIGURE 4

The simplest possible rectangle-to-square dissection is the conversion of a 1×4 rectangle to a square, as shown in Figure 2. If the length-to-width ratio of the rectangle is not exactly 4 to 1, a two-piece dissection using a straight cut is not possible.

Certain other rectangles can be cut into only two pieces and converted into squares if the "step-cut" technique is utilized. A 9×4 rectangle can be cut into two pieces and transformed into a 6×6 square as in Figure 3. The rises in this two-step dissection are equal to 2 units, the treads are equal to 3. A 16×9 rectangle can be transformed into a 12×12 square using a three-step dissection (Figure 4) and a 25×16 rectangle transformed into a 20×20 square using a four-step procedure (Figure 5).

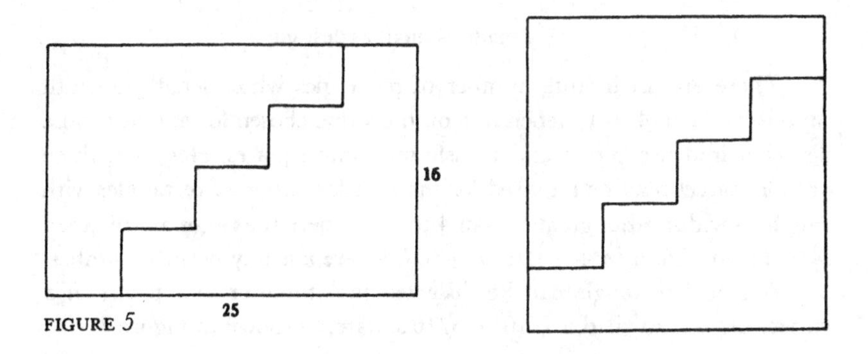

FIGURE 5

From these examples, the following rule can be easily derived: an n-step, two-piece dissection of a rectangle to a square can be made if the dimensions of the rectangle are in the ratio $(n + 1)^2$ to n^2 where this ratio is equal to or less than 4 to 1. The dimensions of the sides of the resulting square will be $n^2 + n$. From this comes the conclusion: as n approaches infinity, the ratio $(n + 1)^2$ to n^2 approaches 1. The 2-piece, nearly infinite-step dissection of such a rectangle into the corresponding 1×1 square is shown in Figure 6. What appears to be a diagonal line in the figure is, of course, a nearly infinite number of steps.

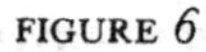
FIGURE 6

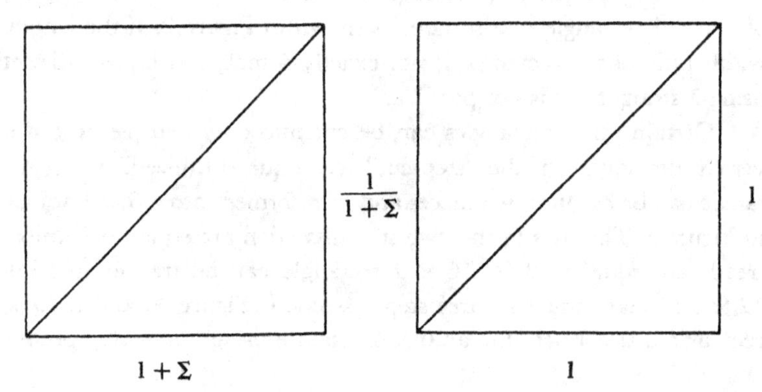

Σ is made as small as desired

There are an infinite number of rectangles whose length-to-width ratio is less than 4 to 1, depending on the value chosen for n, that can be dissected into two pieces and transformed into squares. However, three or more pieces may be required for the transformation of rectangles with length-to-width ratios greater than 4 to 1 or where the ratios are between $(n + 1)^2$ to n^2 and $(n + 2)^2$ to $(n + 1)^2$, where n is any positive number.

A 5×2 rectangle can be dissected into three or four pieces in a variety of ways to yield a $\sqrt{10} \times \sqrt{10}$ square, as shown in Figure 7.

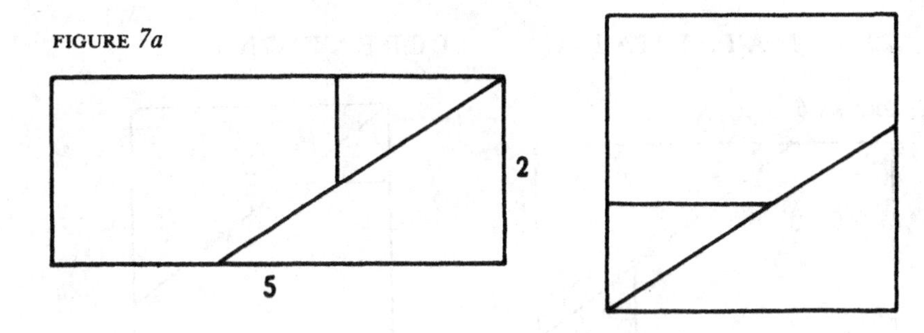
FIGURE 7a
2
5

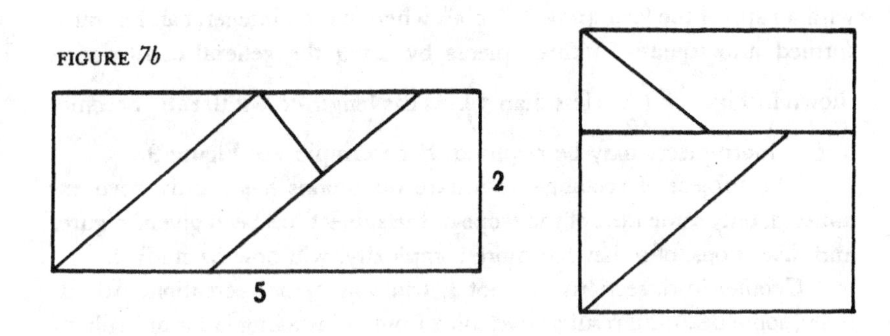
FIGURE 7b
2
5

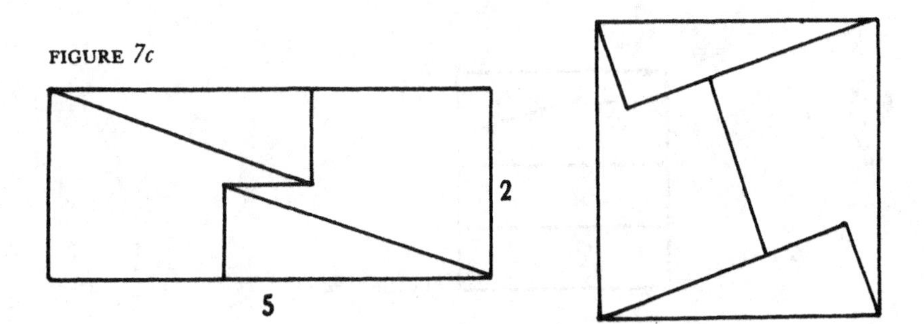
FIGURE 7c
2
5

FIGURE 8

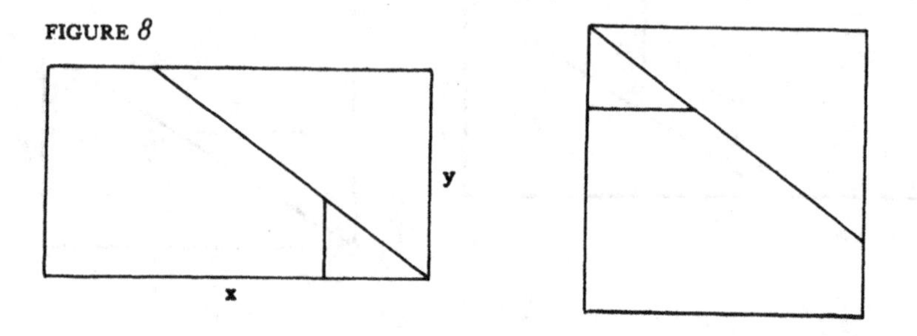

Rectangles whose length-to-width ratios are less than 4 to 1, but not with a ratio of the form $(n + 1)^2$ to n^2, where n is an integer, can be transformed into squares in three pieces by using the general construction shown in Figure 8 $\left(\frac{x}{y} \text{ is less than } 4\right)$. As the length-to-width ratio becomes larger, more pieces may be required. For example, see Figure 9.

The subject of rectangle-to-square dissections has hardly been exhausted, only some idea of the scope of the subject has been given. Figures and dissections, of somewhat more complexity, will now be studied.

Geometric dissections are not a trial-and-error recreation. Admittedly, some beautiful results have come from painstaking labor or brilliant

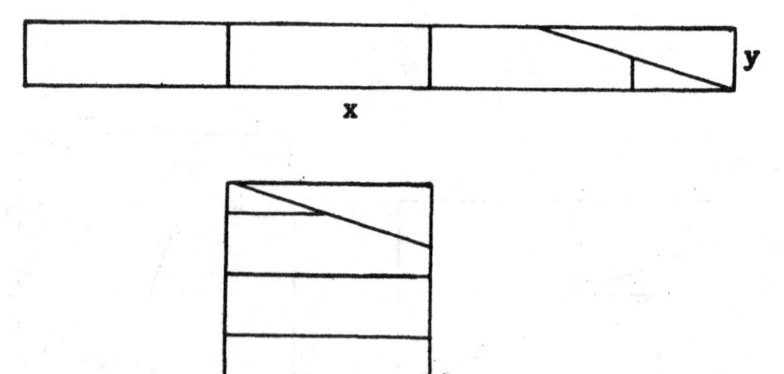

FIGURE 9a

strokes of genius. However, the majority of minimal-piece dissections have resulted from the use of certain standard procedures. One such technique will be illustrated by showing the dissection of an *equilateral triangle,* in which all three sides are equal, into four pieces to form a square.

A strip of squares is drawn (Figure 10*a*), as is a strip of equilateral triangles, each having the same area as one of the squares (Figure 10*b*). If these two strips are superimposed so that the edges of one strip pass through congruent points of the other, the desired dissection can then be constructed. Figure 10*c* shows the method and the result.

More sophisticated methods can yield some marvelous dissections, including curved figures, the conversion of a figure into two or more smaller similar ones, or the combination of several small similar figures into a larger, but different, one. This last example is illustrated by Figure 11, in which a square has been dissected and reassembled to form two pentagons.

Obviously, not all dissections can be solved by using systematic or standard techniques; some problems may require completely novel approaches. Standard techniques usually will point the way toward minimal-piece solutions, but only in a very few exceptional cases can it be proven that a given dissection is indeed minimal. Lindgren has shown that an apparent minimal-piece dissection can often be improved.

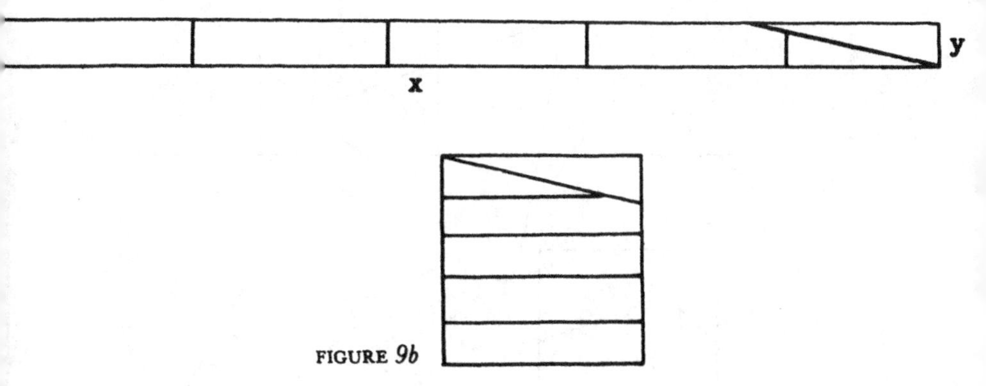

FIGURE *9b*

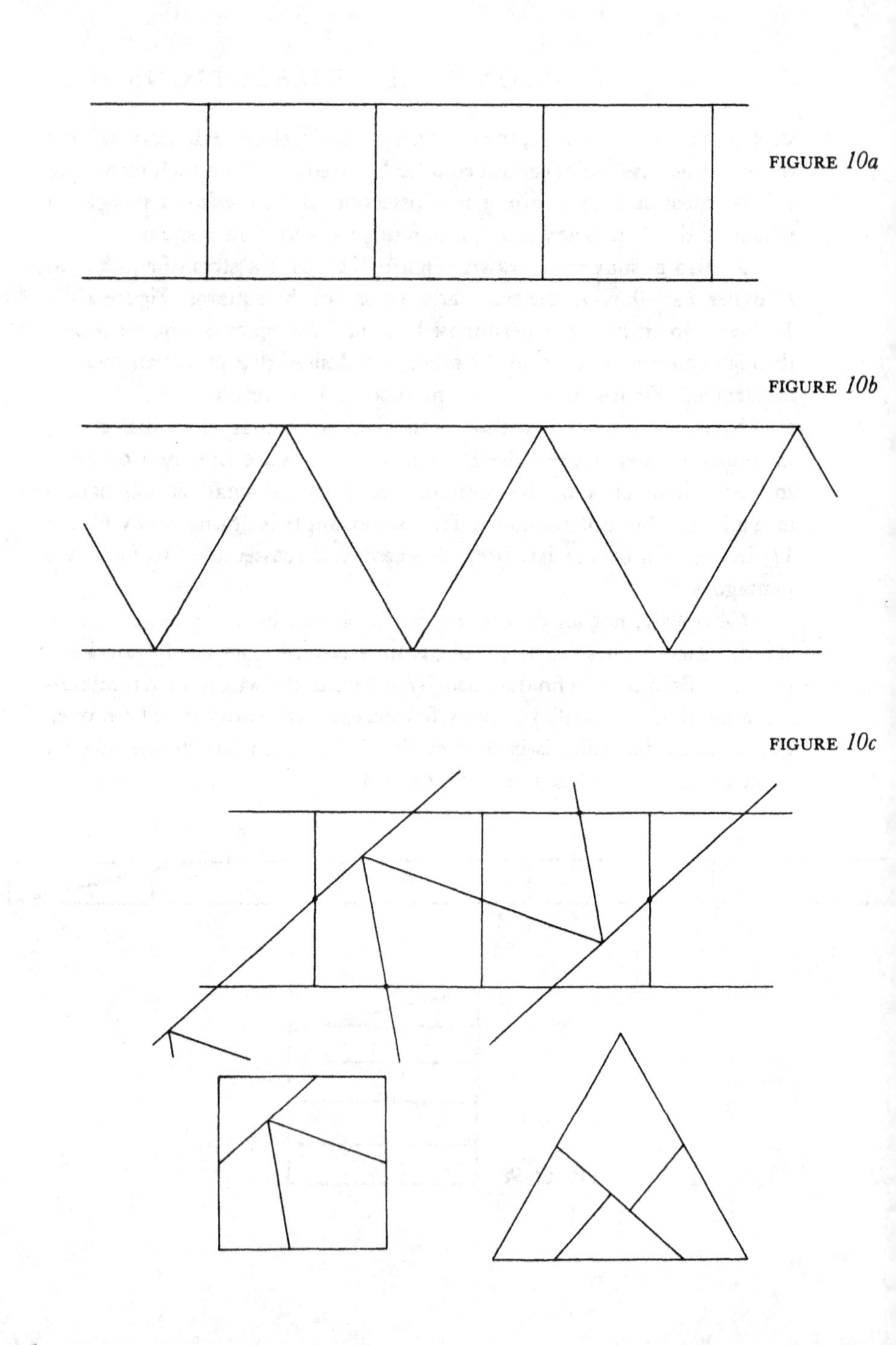
FIGURE 10a
FIGURE 10b
FIGURE 10c

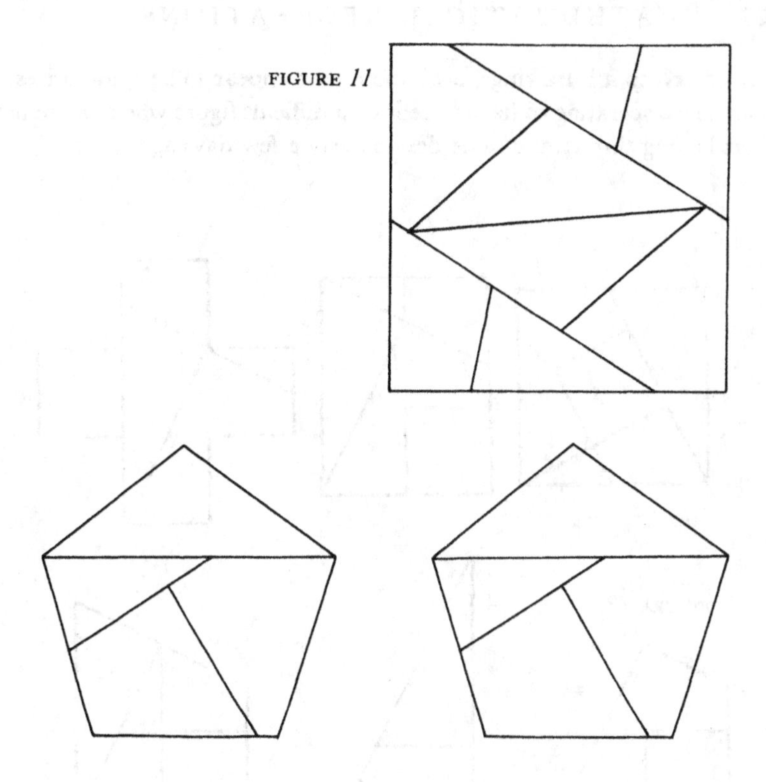
FIGURE *11*

Those who wish to indulge in dissections of their own should follow these hints:

1. Decide on a fixed area for all polygons. Approximately 4 square inches is a convenient area, but any one will do so long as the polygons are large enough that small pieces can be included in the dissection without confusion. The area of each regular polygon will, of course, have to be calculated, and the lengths of the sides of the different polygons will have to be determined.

2. Make accurate black-ink drawings on tracing paper, so that the figure can be turned over and seen in reverse. The lines should not be heavy.

3. Keep all drawings, even those that appear to be quite useless. It can be exasperating to have to redraw a difficult figure when one remembers having thrown the same design away a few days ago.

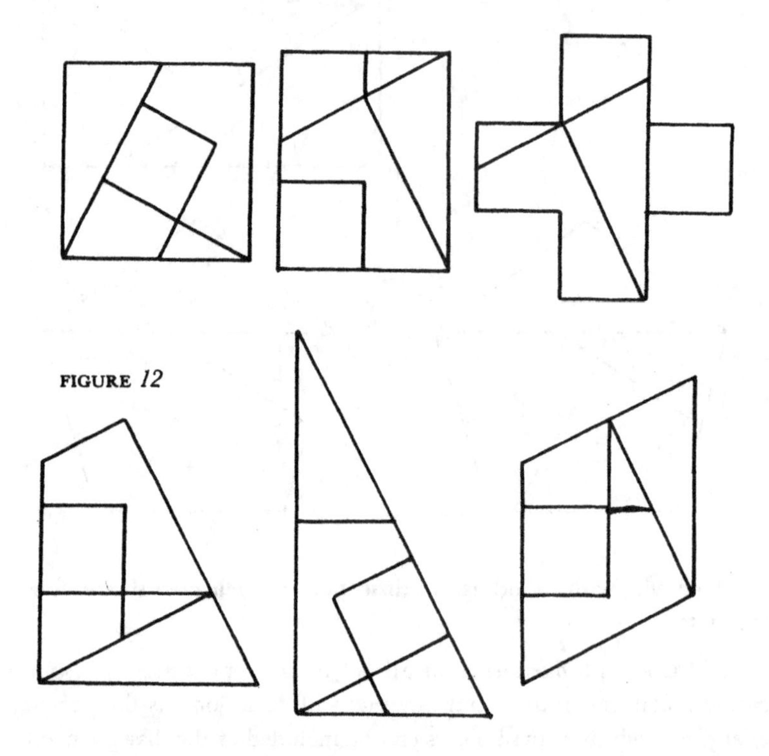

FIGURE *12*

I have indicated that standard dissection techniques usually produce the best results, but nonstandard techniques can also produce marvelous dissections. Sam Loyd's dissection of a square into five pieces that can be rearranged into five other geometric figures is shown in Figure 12.

FIGURE *13*

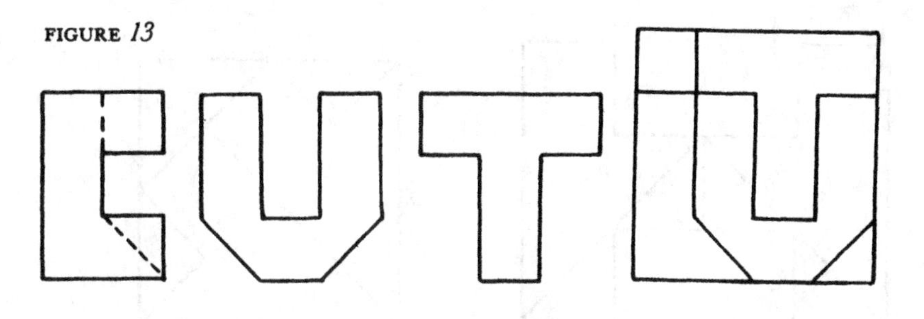

Also, polygons that are not standard in shape can be dissected—for example, the letters of the alphabet. Figure 13 shows how to cut CUT in two places and rearrange the pieces to form a square. The squat H in Figure 14 can be cut into four identical pieces—hardly startling, but it

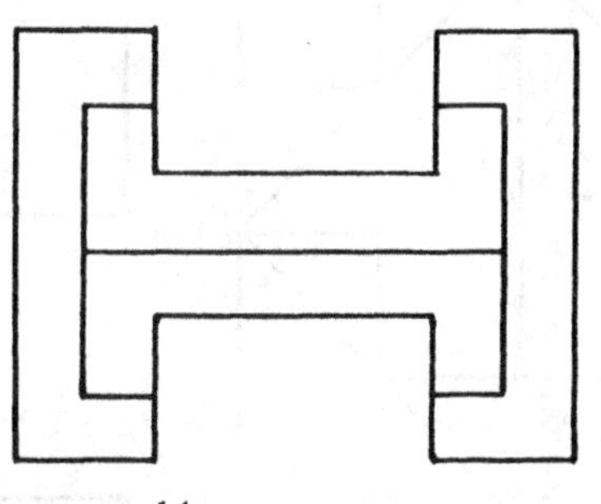

FIGURE *14*

leads to the H^2 problem: Take the same H, make only one cut, and rearrange the pieces to form a square. The solution is shown in Figure 15. The H is first folded along the dotted line. Then a single cut is made, slicing through both sides of the figure. The five pieces that result form the square.

This five-piece dissection is minimal and can be derived from a tessellation technique as shown in Figure 16. A *tessellation* is a tilelike pattern

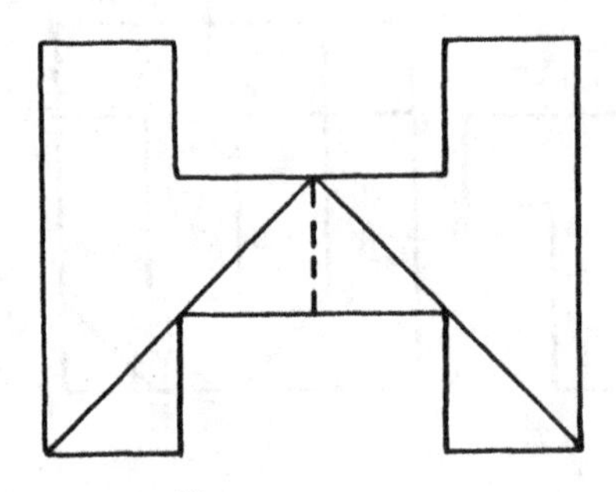

FIGURE *15*

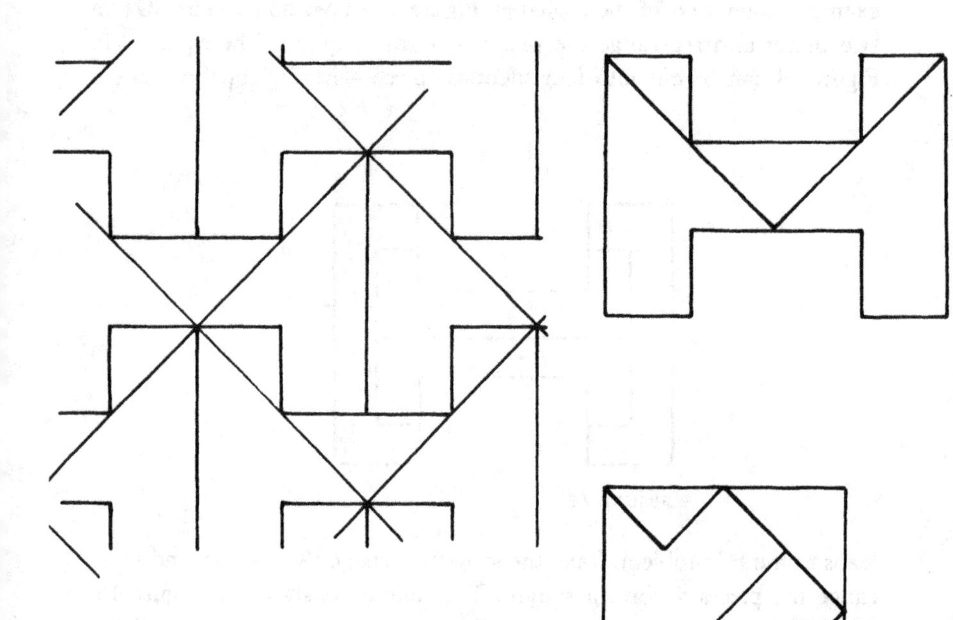

FIGURE *16*

produced on a plane by the repetition of a given figure. In Figure 16, the two tessellation patterns are the repeated squares and the repeated H's. The same method can be used in another five-piece dissection to transform the H into one large Greek cross (Figure 17), or, using six pieces, into two smaller and equal Greek crosses (Figure 18).

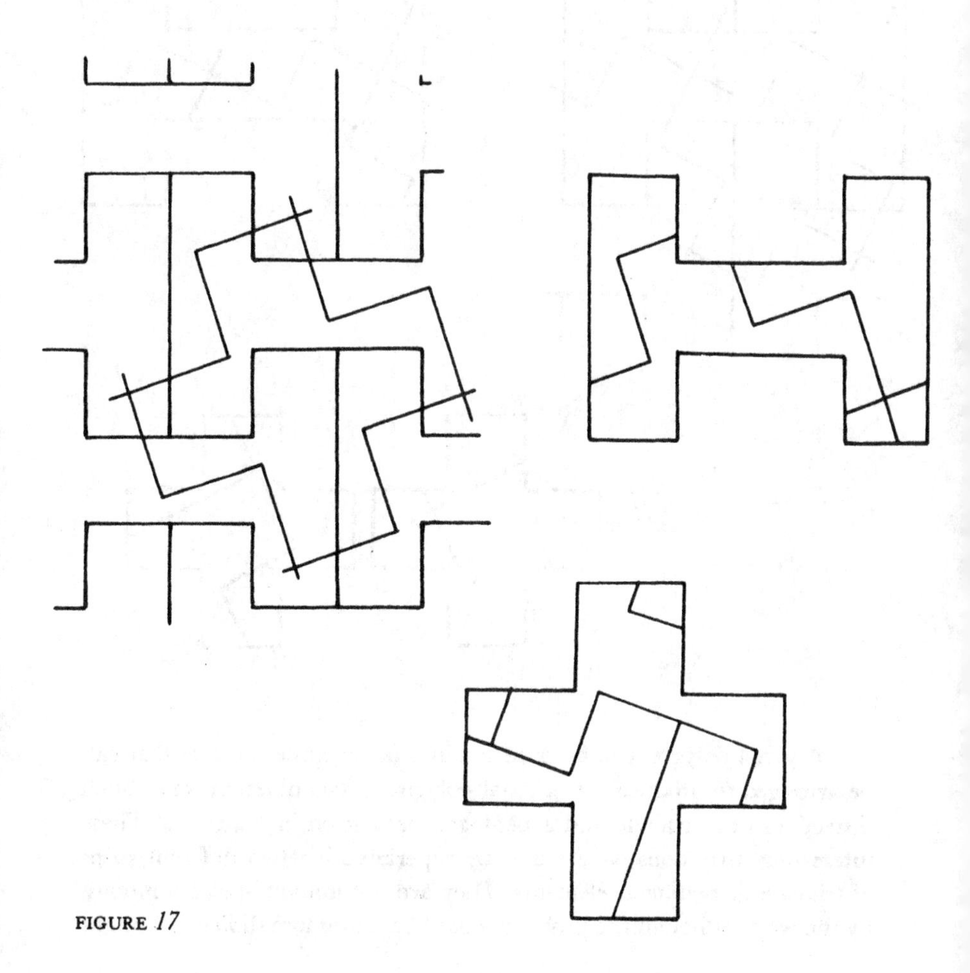

FIGURE *17*

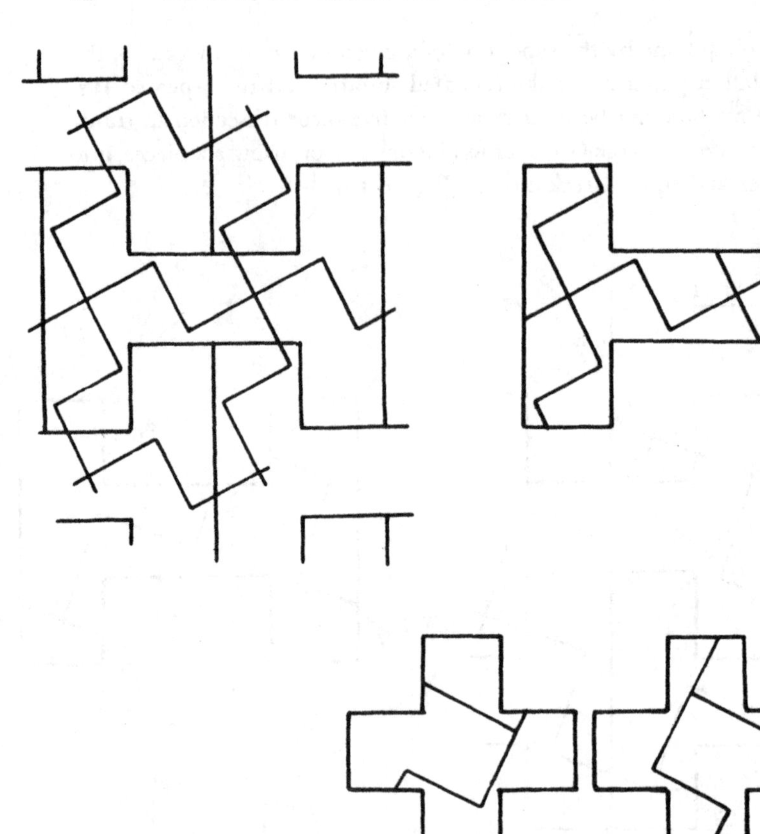

FIGURE *18*

A given polygon can be dissected into noncongruent pieces that can be arranged to produce the original polygon in two different ways. Such dissections of a triangle and a pentagon are shown in Figure 19. These interesting dissections were found by superimposing two different strips of triangle or pentagon elements. They are not minimal-piece solutions, by the way. Although the problem would be a bit more difficult, it might

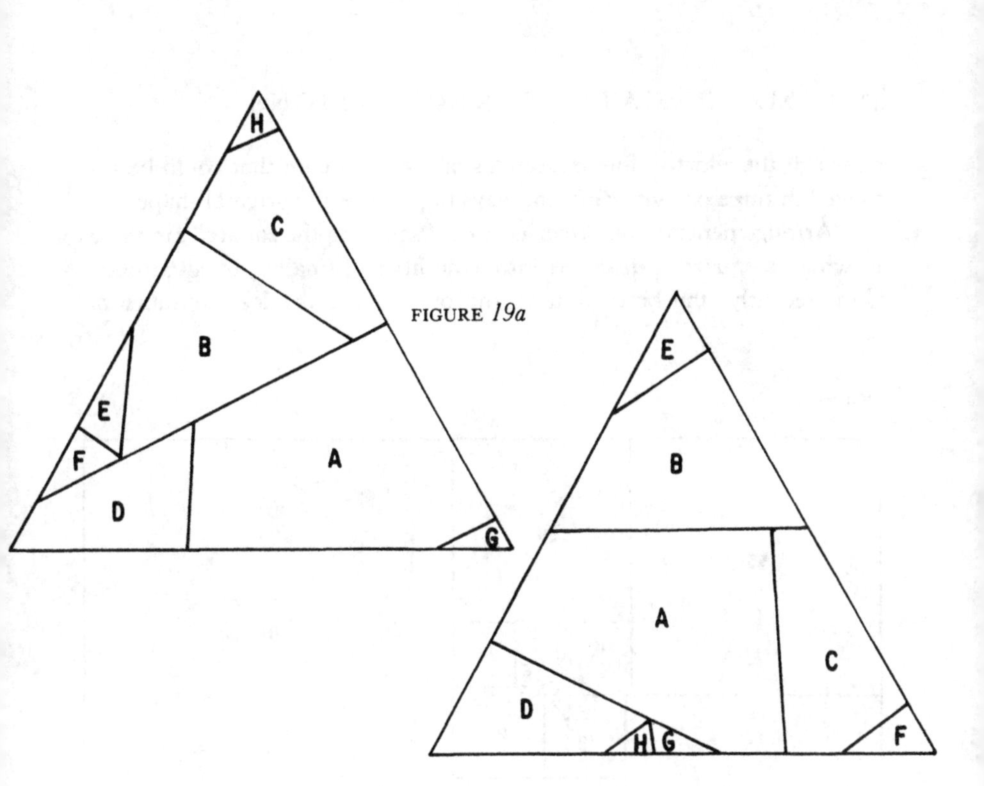

FIGURE *19a*

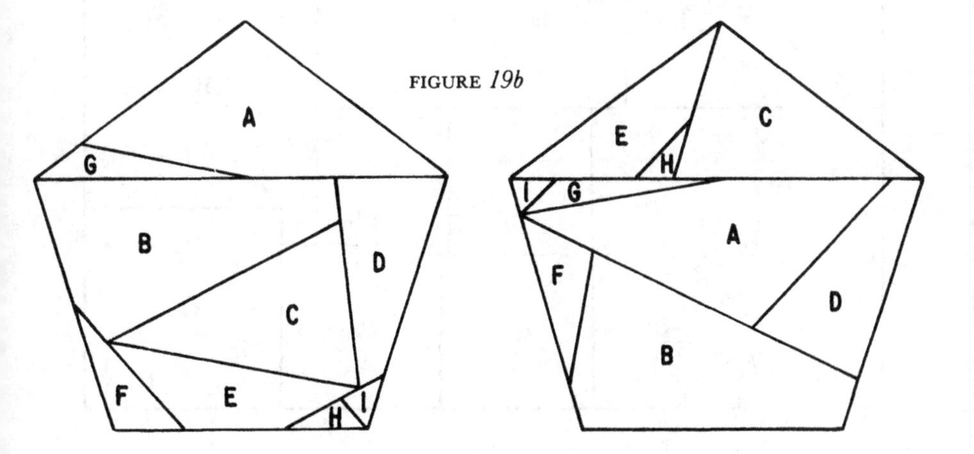

FIGURE *19b*

be worth the effort to find dissections of a given figure that could be arranged in three or more different ways to produce the original shape.

Arrangements usually referred to as "squaring the square" are those in which a square is dissected into a number of smaller, unequal ones. Until recently, the best—that is, the one having the least number of

FIGURE *20*

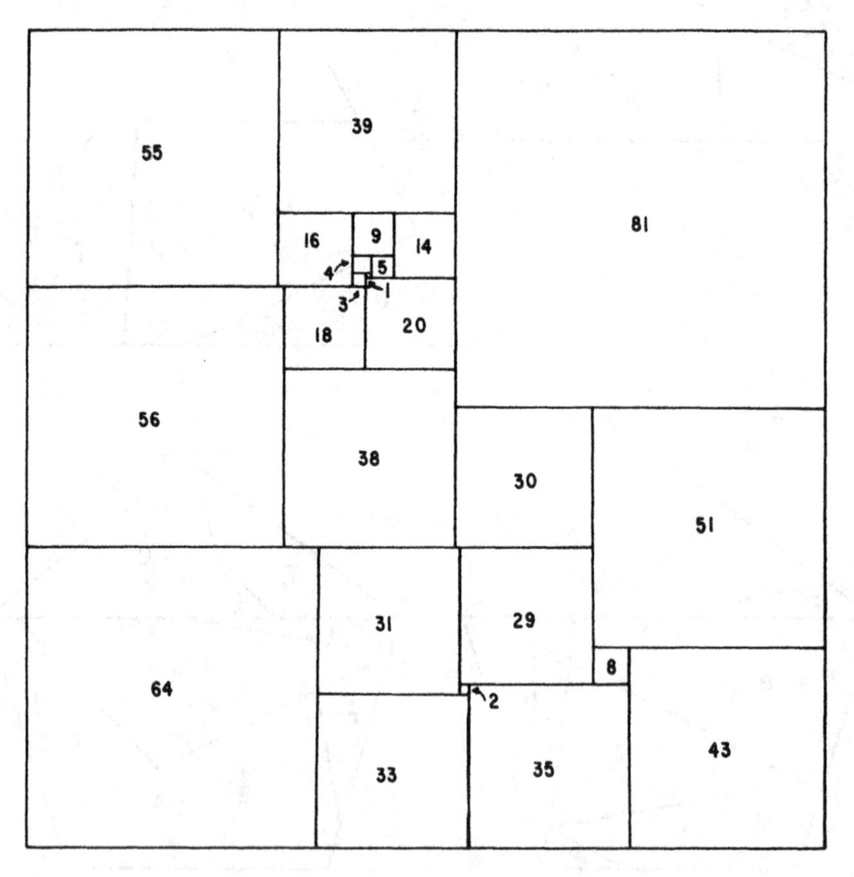

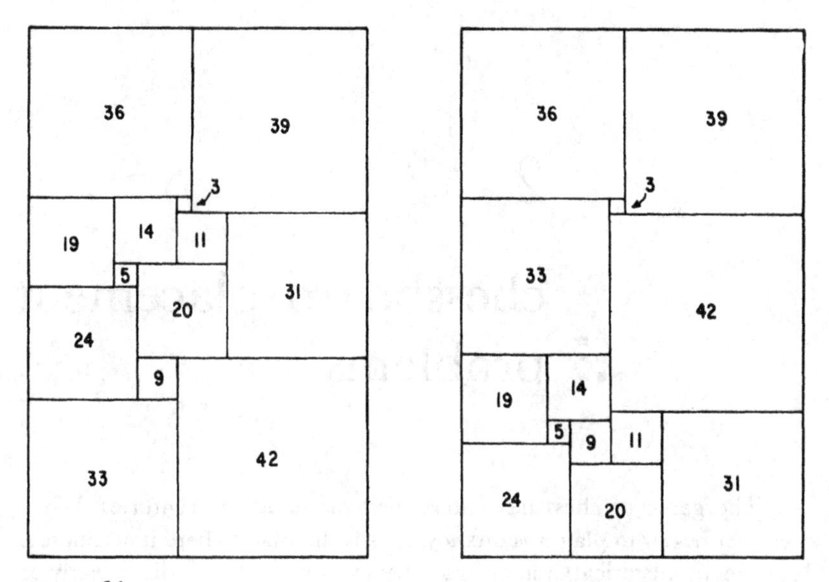

FIGURE *21*

smaller squares—was the one found by an English student of the subject, T. H. Willcocks; it is shown in Figure 20. Each of the numbers gives the lengths of the sides of the squares in which they are placed; the whole square is 175 × 175.

More recently, in 1978, the lowest possible order of perfect square was discovered by A. J. W. Duijvestijn of Twente University of Technology in the Netherlands. Employing a sophisticated computer program Duijvestijn obtained an order 21 perfect square with 112 units on a side. A reproduction of this perfect square can be found on page 86 of the June 1978 issue of *Scientific American*.

One of the many interesting results of the research into squaring the square was the discovery of a rectangle that could be dissected into unequal squares, which could then be arranged in another way to form the original rectangle. This dissection is shown in Figure 21.

2

chessboard placement problems

The game of chess has entertained mankind for centuries. When men first began to play it is unknown, as is the place where it originated. It is general historical opinion that chess was played in India as early as the sixth century A.D. Problems associated with the various pieces and their placement, unrelated to any actual or hypothetical game, have probably entertained puzzlists for an equal length of time.

For example: What is the maximum number of chess pieces of a given type that can be placed on a chessboard, on which no other type of piece is put, so that no piece can be captured by another? Someone may ask, how many white queens can be placed on a board so that no queen can attack another? In this case, the answer is that, since white queens cannot capture white queens, the number that can be placed is limited only by the manual dexterity of the person who piles them on the board. Such questions and answers will be ignored.

Assuming queens to be belligerent toward each other but still limited in their movements to those allowed in chess, we can ask: What is the maximum number of queens that can be placed on a chessboard so that no piece can be captured by another? The queen can move an unlimited number of squares horizontally, vertically, or diagonally. Since a chess-

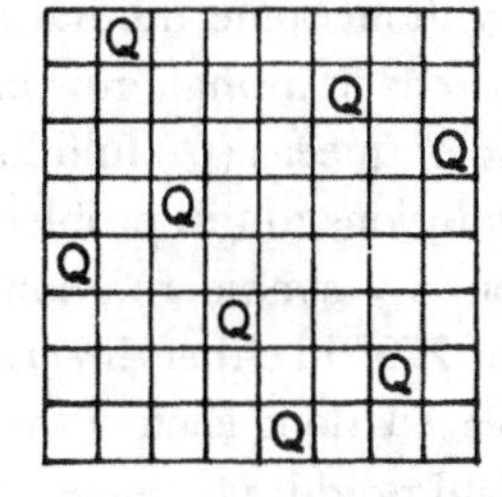

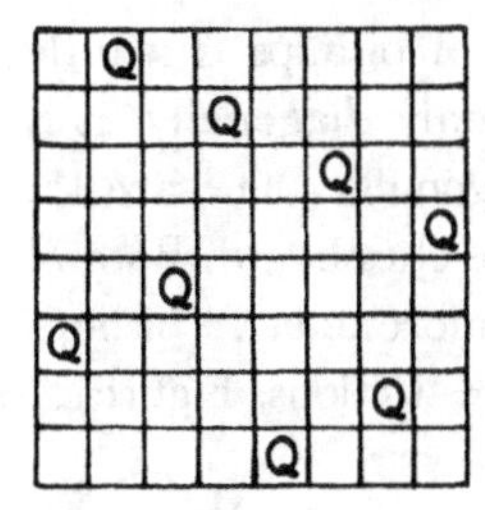

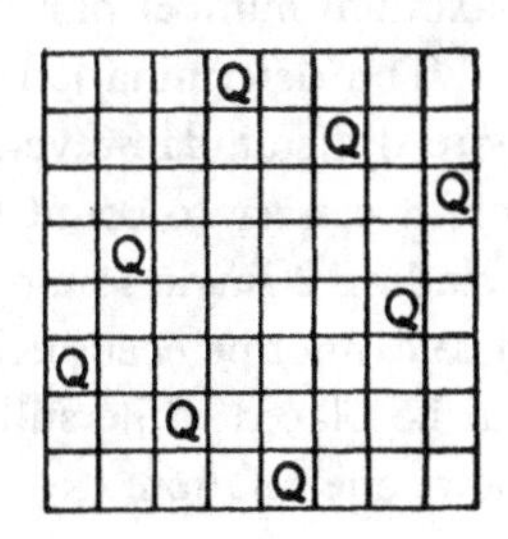

FIGURE *22*

board has eight horizontal rows and eight vertical columns, nine or more queens cannot be placed without some mutual attack, for there would obviously be at least two queens in a single row or column.

The maximum number of queens (Q) fulfilling the stated conditions is 8. There are 12 distinct solutions to this problem, and 80 others can be derived from these solutions by simple rotation or reflection. *Rotation,* turning the board 90, 180, or 270° in either direction, and *reflection,* reflecting the board in a mirror placed along each of the four sides of the board, of the 12 distinct solutions will produce 84 more. The last board in Figure 22 yields only 3 different solutions since only the 90° rotation is different from the original; the 270° rotation duplicates the 90° rotation; the right and left mirror images are identical, as are the top and bottom mirror images. Hence, there is a total of 12 + 80, or 92, solutions to the problem of the queens.

The same question could be asked of other chess pieces. A rook (R) can move any number of squares horizontally or vertically, but it cannot move diagonally; a bishop (B) can move any number of squares diagonally; a knight (Kt) can move only from one corner of a 2×3 rectangle to the corner diagonally opposite; a king (Kg) can move only one square horizontally, vertically, or diagonally.

The maximum number of rooks that can be placed on a chessboard without any one capturing another can be calculated readily. Eight rooks could be placed along one of the main diagonals, and no mutual capture would be possible (see Figure 23). A ninth rook would necessarily have to be placed in a row or a column that was already occupied. Hence, the maximum number of rooks is 8.

The determination of the maximum number of bishops is a little more difficult. However, since bishops can move only diagonally, it is only necessary to count the number of parallel diagonals. There are 15, including 2 single squares on opposite corners of the chessboard. Both of these cannot be occupied simultaneously, hence no more than 14 bishops can be placed while still conforming to the given restrictions. Figure 24 shows one solution.

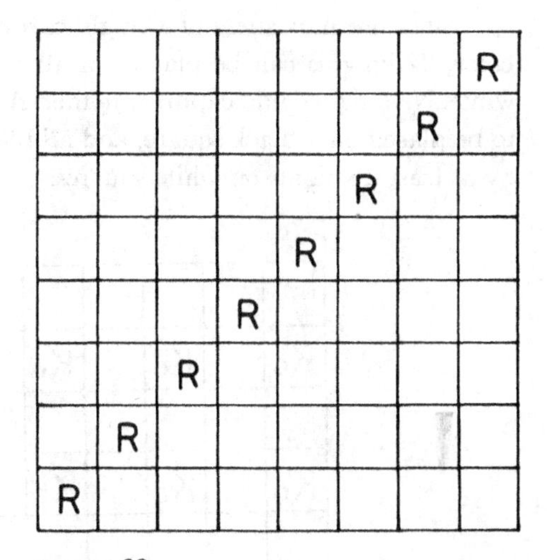

FIGURE *23*

FIGURE *24*

B							
B							B
B							B
B							B
B							B
B							B
B							B
B							

Since one movement of a knight carries him to a square of a different color, 32 knights can be placed on all 32 squares of a given color, say white. No knight could capture another. A thirty-third knight would have to be placed on a black square, and all black squares are liable to attack by at least 2 knights on white squares.

KG		KG		KG		KG	
KG		KG		KG		KG	
KG		KG		KG		KG	
KG		KG		KG		KG	

FIGURE *25*

Kings must be separated by at least one square, and the best arrangement, every other square, shows that no more than 16 kings can be so placed (see Figure 25).

Determining the maximum number of pieces that can be placed on boards larger than the 8×8 chessboard follows the same line of reasoning applied to the standard board. For any $n \times n$ board, the maximum number of queens or rooks is n. On an $n \times n$ board there are $2n - 1$ parallel diagonals, and both of the squares at opposite ends of the longest diagonal cannot be occupied. Therefore, no more than $[(2n - 1) - 1]$, or $2n - 2$, diagonals can have bishops. If n is even and >2, $\frac{n^2}{2}$ knights can be placed, since just that many squares of each color are on the board. If n is odd, then there is one square more of one color than the other; that is, $\frac{(n^2 + 1)}{2}$ squares of one color and $\frac{(n^2 - 1)}{2}$ of the other. The maximum number of knights that can be placed on an $n \times n$ board when n is odd and >1 is $\frac{(n^2 + 1)}{2}$. The every other square arrangement for kings applies to all boards. It is apparent, however, if Figure 25 is examined, that 16 kings can be placed on a 7×7 board as well as on the 8×8 one. Hence, the maximum number of kings that can be placed on an $n \times n$ board is dependent on whether n is odd or even. If n is odd, then $\frac{(n + 1)^2}{4}$ kings can be placed; if n is even, then $\frac{n^2}{4}$ kings can be placed.

A variation of this type of problem is to determine the minimum number of pieces of a given type that can challenge, or legally attack, according to the usual chess rules for moving the pieces, or occupy all the squares of a chessboard when the restriction against mutual capture is ignored. Figure 26 shows a solution to this problem using queens. The

FIGURE *26*

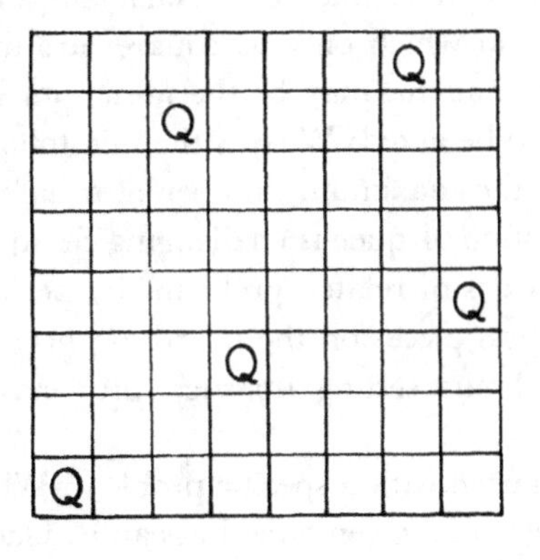

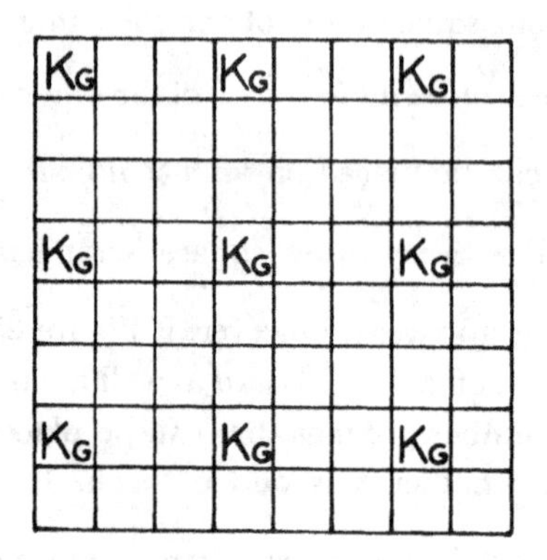

FIGURE *27*

minimum number of queens is 5. Figure 27 shows that only 9 kings are needed to challenge or occupy all 64 squares. The corresponding minimum-piece solutions for the problem using bishops, rooks, and knights are shown in Figure 28.

Another variation of the placement problem is to determine the minimum number of squares that can be challenged by 8 queens. Figure 29 shows a solution in which only 53 squares are under attack by this number. It appears that this may be the minimum solution. The opposite question can also be asked: What is the minimum number of queens that will challenge the maximum number of squares? Figure 30 shows the best known solution, 4 queens challenging 62 squares. It is a simple matter to devise dozens of related problems by permitting two or more different kinds of chess pieces on the board at once, and the reader can while away many hours setting up such problems and finding their solutions.

This chapter started with a specific problem: What is the maximum number of chess pieces of a given type that can be placed on a chessboard so that no one can be captured by another? The question was answered

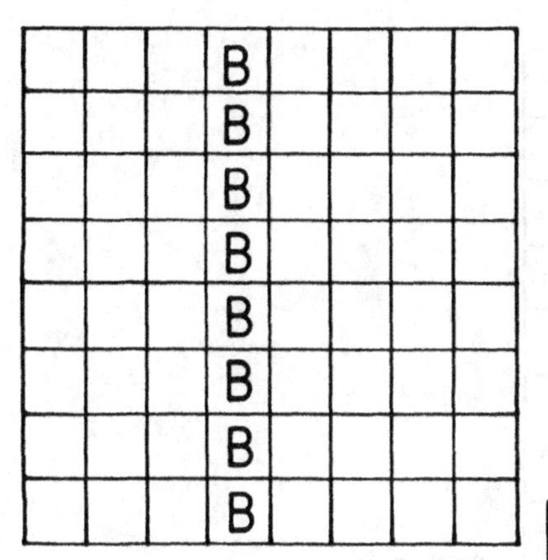

FIGURE *28a*

FIGURE *28b*

							R
						R	
					R		
				R			
			R				
		R					
	R						
R							

FIGURE *28c*

					Kt		
	Kt	Kt		Kt	Kt		
		Kt					
					Kt		
		Kt	Kt		Kt	Kt	
		Kt					

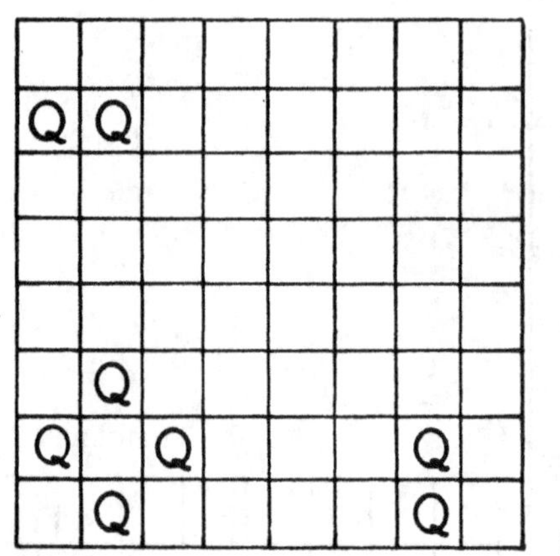

FIGURE *29*

for various pieces, but how many different solutions are there for each given chessman?

As has been noted, there are 92 solutions for the 8-queen problem, and they are all derivable from the 12 distinct answers given in Figure 22. For larger boards, the number of solutions for the queen problem increases quite rapidly. For a 9×9 board, the 9 queens can be arranged in 352

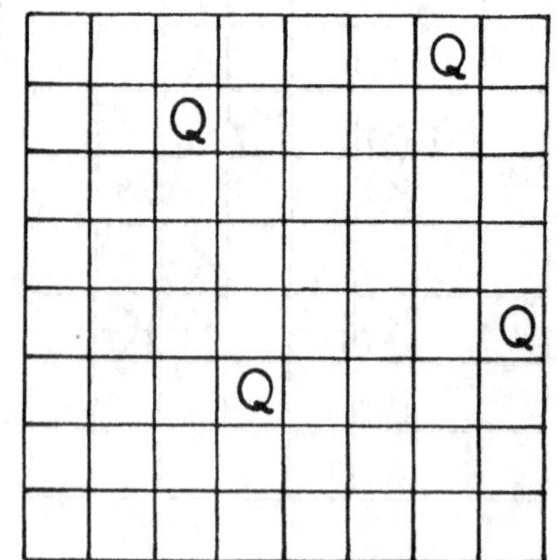

FIGURE *30*

ways, of which 46 are distinct; 10 queens on a 10×10 board can be arranged in 724 ways, of which 92 are distinct; 11 queens on an 11×11 board can be arranged in 2,680 ways, of which 341 are distinct. The results are known up to the 12×12 board, for which the solution was established mathematically by J. W. L. Glaisher, an English mathematician, in 1874. For this board, there are 14,200 ways of arranging the 12 queens, and 1,784 of these are distinct solutions.

The 32 knights can occupy either color square, so there are but 2 distinct solutions for the knight problem. The 16 kings can be arranged in 16 different ways. For the 14 bishops, there are 256 solutions, but many are merely reflections or rotations of a number of fundamental ones. Dudeney found that there were only 36 distinct solutions to the bishop problem. The method of establishing these 36 distinct solutions, without actually constructing all 256 and making comparisons, will be given here as a preparation for the outline of the solution to the 8-rook problem which, until recently, was unknown.

It has already been established that the maximum number of bishops that can be placed on an $n \times n$ board is $2n - 2$. The total number of *permutations,* or ways of arranging these pieces, is 2^n. Thus, there are 256 permutations for the 8 bishops on a chessboard. It will now be established that only 36 of these solutions are distinct.

Two, and only 2, corner squares must be occupied. All bishops, in order to challenge the maximum number of squares simultaneously, must be placed on border squares, that is, on squares at the edge of the board. There are two classes of solution, as shall be seen, which correspond to 4 and 8 permutations each.

Since 2 corner squares not along the same diagonal must be occupied and the other 2 left vacant, the 256 permutations can be divided by 4, and only those arrangements in which the 2 corner squares at the bottom are occupied need be considered. Rotation of the board would place these squares along either side or the top or bottom. Since these are really only different orientations of the same arrangement, only one of them need be considered. There now are $^{256}/_{4}$, or 64, permutations to study for possible reflection or rotation solutions.

There are 12 bishops not on corner squares, and they are in pairs across the board as illustrated in Figure 31. Bishop 1 can be placed in 1*A* or 1*B*, but wherever it goes, bishop 2 must go in a corresponding square, 2*A* or 2*B*, to avoid attacking it. The same holds true for bishops 3, 4, and so forth, so there are 2 choices for each of 6 pairs, which represent the 2^6, or 64, permutations.

For solutions that are horizontally symmetrical—that is, in which the bishops on the left side of the board are mirror images of those on the right side—the placement of bishops 1*A* and 1*B* determine the placement of bishops 2*A* and 2*B* and, therefore, of bishops 11*A* and 11*B* and 12*A* and 12*B* as well. Thus, there are 2 choices for each of 3 sets of 4 bishops, or 8 horizontally symmetrical permutations, making 8 distinct solutions.

The remaining 56 permutations are those that are not horizontally symmetrical; for one such solution see Figure 32. There is a mirror-image

-	12B	10B	8B	6B	4B	2B	-
11A							2A
9A							4A
7A							6A
5A							8A
3A							10A
1A							12A
(B)	1B	3B	5B	7B	9B	11B	(B)

FIGURE *31*

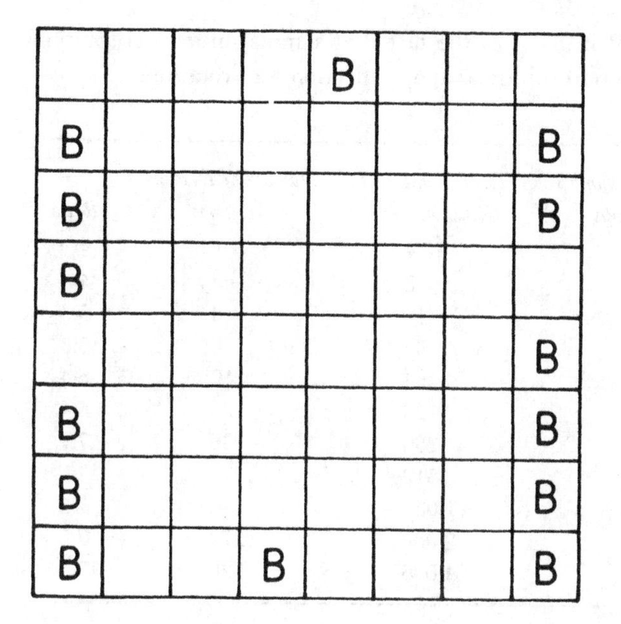

FIGURE *32*

solution as well. There are 28 distinct solutions that are not horizontally symmetrical. This brings the total of distinct solutions to exactly 36.

Dudeney also determined the number of distinct solutions to the bishop problem for boards of any size. The number of distinct solutions on an $n \times n$ board is $2^{\frac{(n-4)}{2}}(2^{\frac{(n-2)}{2}} + 1)$ if n is even, and $2^{\frac{(n-3)}{2}}(2^{\frac{(n-3)}{2}} + 1)$ if n is odd. The table on page 46 lists the data for the bishop problem for boards from 2×2 to 12×12. The last column in the table is the ratio of the total number of solutions to the number of distinct solutions. Since there are 4 mirror image and 4 rotation solutions to each distinct answer, it is obvious that there can never be more than 8 times as many total solutions as there are distinct ones. Dividing 2^n by either of the two formulas given above also shows that, after $n = 3$, this particular ratio of total solutions to distinct solutions approaches, but never reaches, 8. This is to be expected if thought is given to the fact that some symmetrical solutions will not give 8 different solutions upon reflection and rotation.

n	*Maximum Number of Bishops*	*Total Number of Solutions: 2^n*	*Number of Distinct Solutions*	*Ratio*
2	2	4	1	4
3	4	8	2	4
4	6	16	3	5.3
5	8	32	6	5.3
6	10	64	10	6.4
7	12	128	20	6.4
8	14	256	36	7.1
9	16	512	72	7.1
10	18	1,024	136	7.5
11	20	2,048	272	7.5
12	22	4,096	528	7.8

The 8-rook problem is considerably more difficult to analyze than the 14-bishop problem. Even Dudeney confessed he was unable to determine the number of distinct solutions to the rook problem on a chessboard. For an $n \times n$ board, a maximum of n rooks can be placed so that no rook attacks another. For the smaller boards, the problem of deter-

FIGURE 33

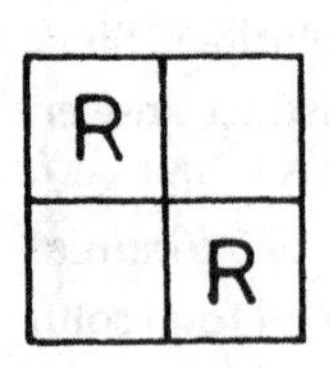

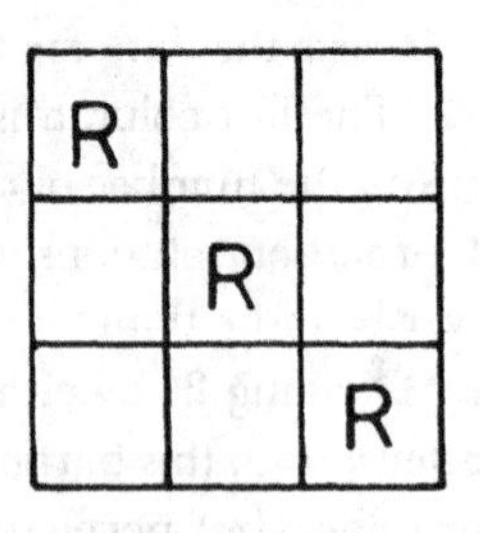

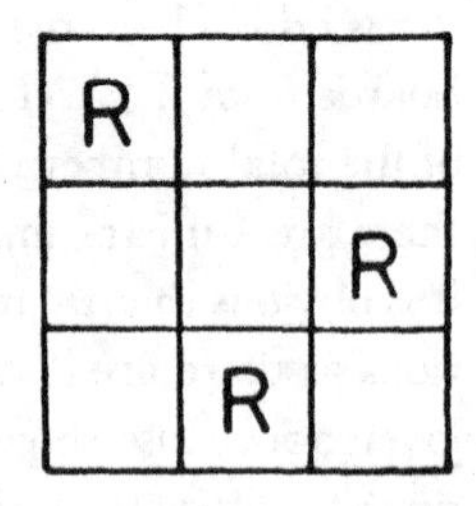

mining all the distinct solutions is relatively simple. Figure 33 shows the single solution for the 2×2 board and the two distinct solutions for the 3×3 board. The total number of solutions for any $n \times n$ board is $n!$, that is, *n factorial,* which means $(1)(2)(3) \cdots (n)$. A 4×4 board has 4!, or (1)(2)(3)(4), or 24, solutions of which 7 are distinct. These are given in Figure 34.

It was shown that the ratio of total solutions to distinct ones is always less than 8 for the bishop problem. The same ratio limit applies to the

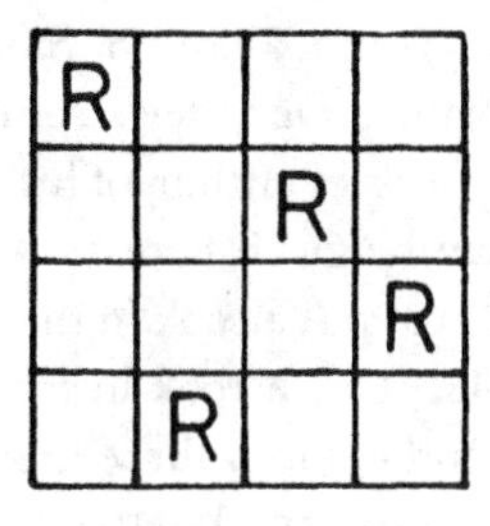

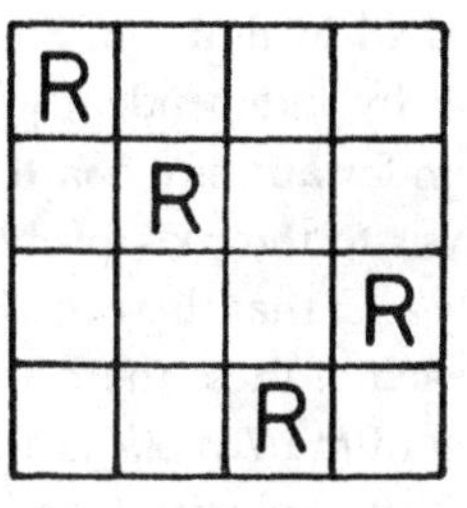

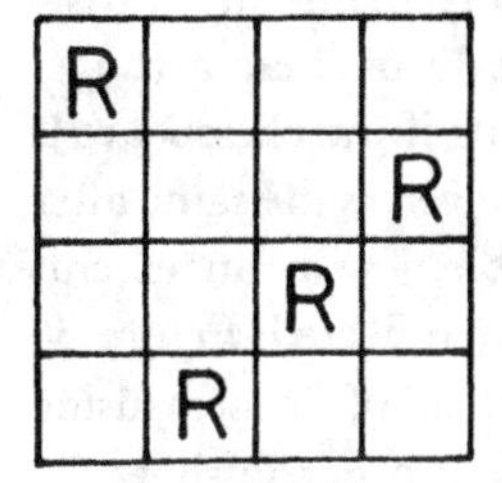

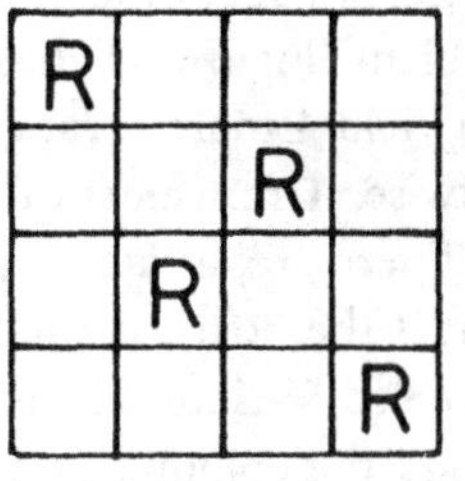

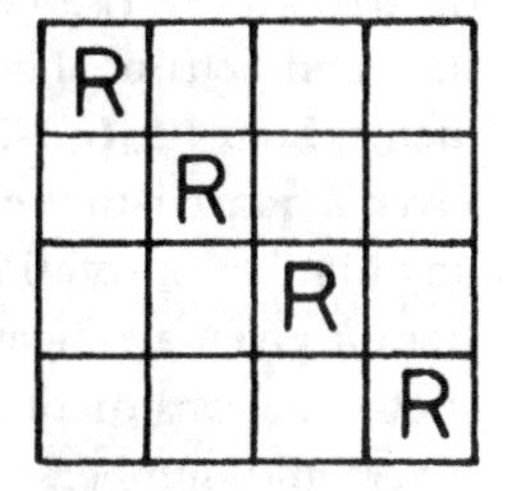

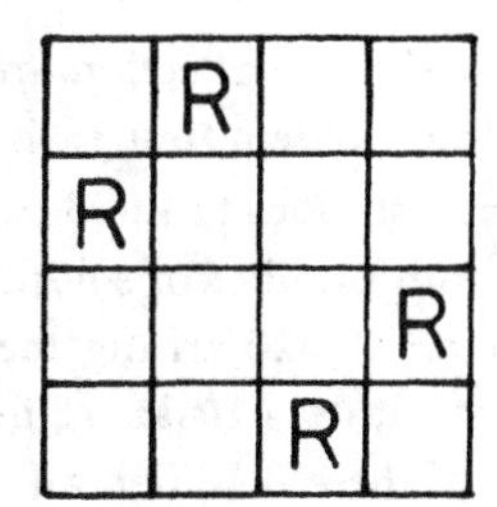

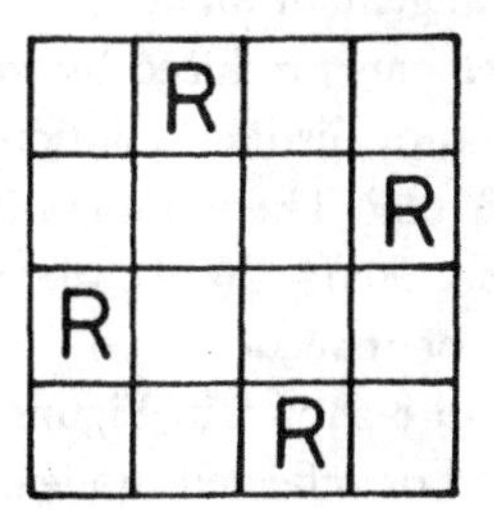

FIGURE *34*

rook problem. However, until recently, the exact number of distinct solutions to this problem was not known. For a 5 × 5 board with 120 possible solutions, it would be a matter of comparatively little time to determine the number of distinct solutions by actual construction and examination. Since there are 8!, or 40,320, possible solutions to the 8-rook problem, actual construction and examination are hardly useful methods. The task of separating rotated or reflected solutions from such an overwhelming number of constructions, if someone were actually to draw them, would be formidable. Late in 1962, the issue was finally attacked in a systematic manner by David F. Smith of San Francisco, California. Smith warmed up to the task by independently confirming Dudeney's solutions to the bishop problem for any size board. Only a brief outline of his rather complicated analysis to the rook problem can be given here.

It is easily seen that there are 8 places to put a rook in the first column and that for each of these there are 7 places for a rook in the second column. For each of the 7 rooks in the second column there are 6 places for a rook in the third column. Continuing across the board, there is a total of (8)(7)(6)(5)(4)(3)(2)(1), or 40,320, different arrangements that satisfy the conditions of the problem. Figure 35 illustrates 5 of these arrangements. Four (*A*, *C*, *D*, *E*) yield 4 more arrangements if the chessboard is given a quarter-turn clockwise. Arrangement *B*, however, is the same after any number of rotations. There are, in fact, 7 possible new arrangements from a given arrangement if the board is rotated or reflected. Figure 36 shows one construction (labeled Variation 1), and the set of 7 others, using letters and numbers to mark the columns and rows in order to show the type of rearrangement made.

Arrangements produced by rotation and/or reflection are not essentially different, or distinct, solutions to the problem, but each is included in the 40,320 total. The number of *distinct* solutions to the 8-rook problem must be determined—that is, the minimum number of solutions that, by rotation and/or reflection, will generate all 40,320 arrangements.

The set of 8 shown in Figure 36 constitute a *closed set;* no other one can be rotated or reflected to yield any of these 8. When all 8 variations of an arrangement are different, as is true for Figure 35*A*, the set can be considered to be one solution corresponding to 8 variations. Such solutions will be termed Class *A* solutions. (Class names will be assigned to match the letters assigned to the patterns in Figure 35.) As the interest here is only in evaluating the number of distinct solutions, it is not necessary to decide which member of a group is to be considered the distinct one, since any others would be considered mere arrangements.

There can be no more than 8 arrangements for each distinct solution, so that a lower limit can be set on the number:

$$40{,}320 \div 8 = 5{,}040$$

FIGURE 35

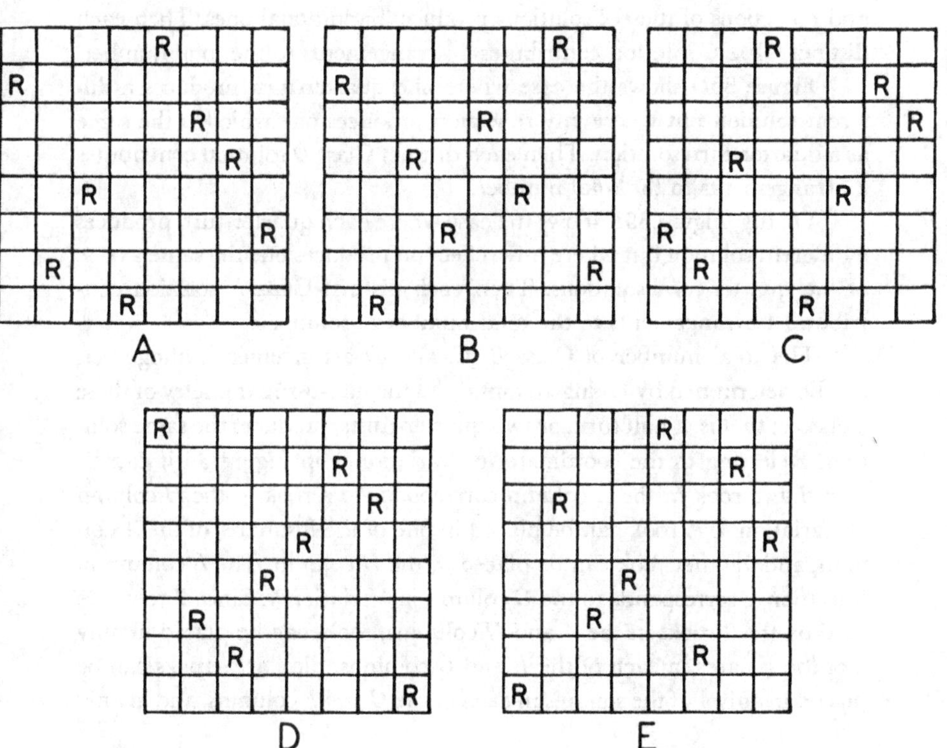

There will be more than this minimum number of distinct solutions, though, because in some cases certain manipulations (rotations or reflections) may yield the original pattern. The number of solutions and their arrangements covered by those cases where all 8 variations are *not* different must be found. Dividing the total of Class *A* variations by 8 will give the remaining number of solutions.

Figure 35*B* shows a solution where some of the 8 variations are the same. In this case any number of quarter-turns produces the same arrangement and any reflection produces only one different solution. So each distinct Class *B* solution contributes two arrangements to the total number of solutions.

In Figure 35*C*, only one quarter-turn produces a different solution, and reflections of these 2 solutions produce 2 additional ones. Then each distinct Class *C* solution contributes 4 arrangements to the total number.

Figure 35*D* shows the case where one quarter-turn produces a different solution but where any reflection produces one which is the same as a quarter-turn solution. Then each distinct Class *D* solution contributes 2 arrangements to the total number.

Finally, Figure 35*E* shows the case where each quarter-turn produces a different solution but where any reflection produces one the same as one of the quarter-turn solutions. Then each distinct Class *E* solution contributes 4 arrangements to the total number of solutions.

The total number of Class *B*, *C*, and *D* arrangements, altogether, can be determined by taking advantage of the half-turn symmetry of these 3 classes; that is, a half turn, or two quarter-turns, produces the same solution. Referring to the coordinates of Variation 1 in Figure 36 it can be seen that a rook in the *A* column corresponds to a rook in the *H* column of Variation 4. A rook can be placed in one of the 8 squares of the *A* column, and another rook can be placed in the *H* column. The *B* column in Variation 1 corresponds to the *G* column in Variation 4. Since 2 rows are used by the 2 rooks in the *A* and *H* columns, rooks can be placed in only 6 of the squares in each of the *B* and *G* columns. Similarly, rooks can be placed in only 4 of the squares in each of the *C* and *F* columns, and in only

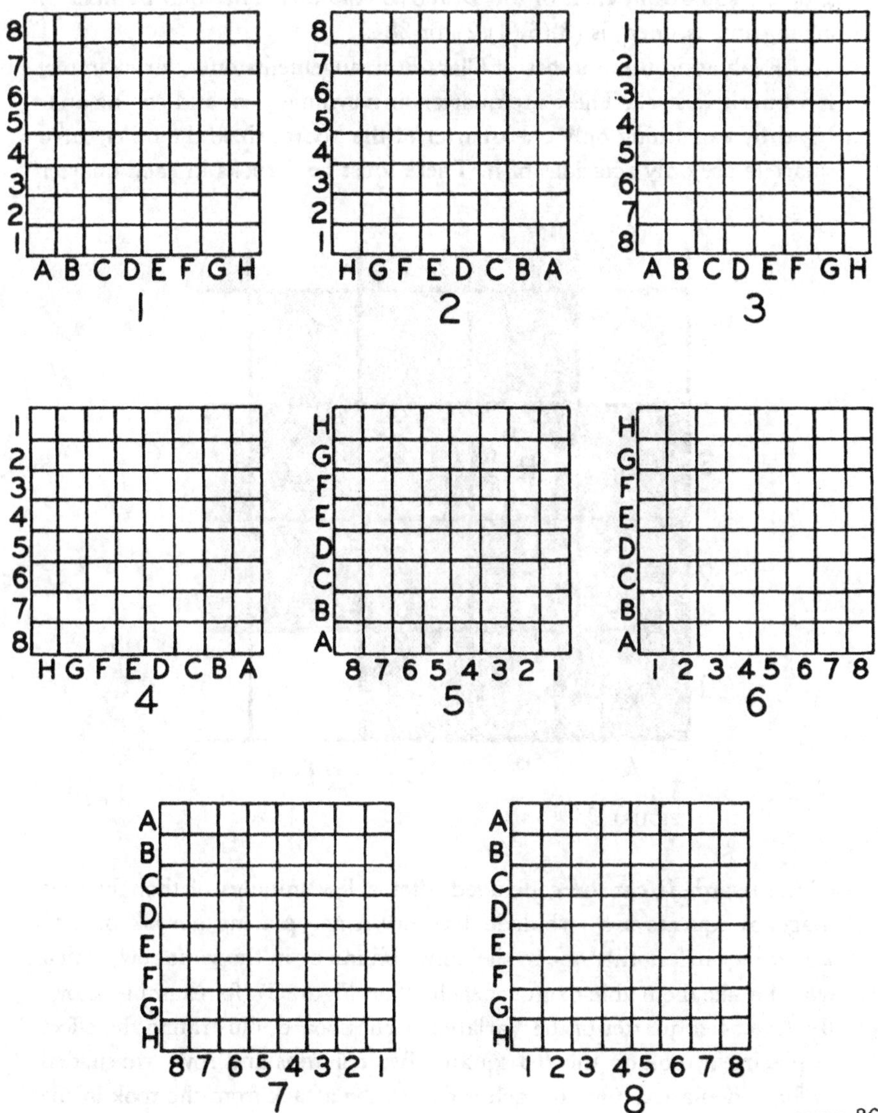
8 7 6 5 4 3 2 1
A B C D E F G H
1
8 7 6 5 4 3 2 1
H G F E D C B A
2
1 2 3 4 5 6 7 8
A B C D E F G H
3
1 2 3 4 5 6 7 8
H G F E D C B A
4
H G F E D C B A
8 7 6 5 4 3 2 1
5
H G F E D C B A
1 2 3 4 5 6 7 8
6
A B C D E F G H
8 7 6 5 4 3 2 1
7
A B C D E F G H
1 2 3 4 5 6 7 8
8

FIGURE *36*

2 of the squares in each of the *D* and *E* columns. The total number of arrangements, then, is (8)(6)(4)(2), or 384.

Evaluating the number of Class *B* arrangements only, for example, is a puzzle in itself. These are quarter-turn symmetrical and can be analyzed by examining only one quarter of the board, since the other three quarters are only rotations of it. There must be 2 rooks in each quarter

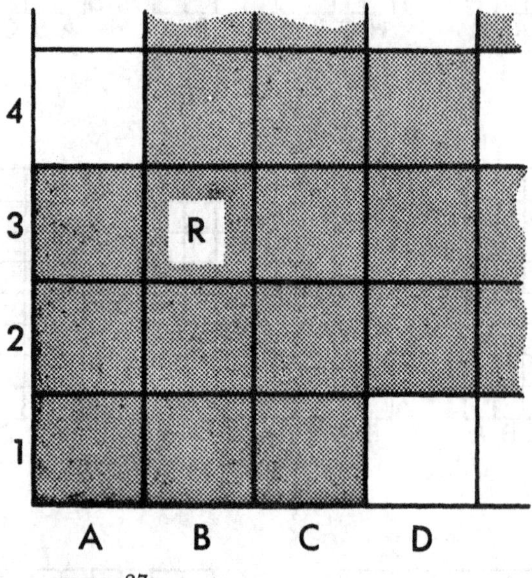

FIGURE 37

of the board. It can be concluded after a few minutes of thought that diagonal squares are forbidden territory, since placing a rook on one would require another one in the same column or in the same row, which would make them able to attack each other. Figure 37, for example, shows the lower-left quarter of the Variation 1 chessboard, illustrating the effect of placing a rook on the *B* 3 square. Two columns and rows are shaded as forbidden areas. One of each is due to the attack from the rook in the position shown; one row is due to the rook's rotation to the lower-right quarter of the board; and a column is due to the rook's rotation to the upper-left quarter of the board. The remaining squares on the diagonal are also shaded.

There are 12 places to put the first rook—16 squares in the quarter of the board, minus the 4 diagonal squares. For each of these choices there will be 2 places not on the diagonal in which to place the second rook. For the *B* 3 placement of the first rook, only squares *A* 4 or *D* 1 can be used by the second rook. There are, then, 24 combinations of the 2 rooks when taken in some specified order and 12 if order is disregarded. Thus, there are 12 Class *B* arrangements. They correspond to 6 distinct solutions.

A similar, but considerably more involved analysis by Smith shows that there are 296 Class *C* arrangements with 74 distinct solutions; 76 Class *D* arrangements with 38 distinct solutions; and 1,376 Class *E* arrangements with 344 distinct solutions.

The total of Class *B*, *C*, *D*, and *E* solutions is 1,760. The number of Class *A* solutions can be determined by subtracting 1,760 from the overall total of 40,320 and dividing this result by eight, giving 4,820. The complete results are:

Class *A* distinct solutions	4,820
Class *B* distinct solutions	6
Class *C* distinct solutions	74
Class *D* distinct solutions	38
Class *E* distinct solutions	344
Total number of distinct solutions	5,282

This number is very close to the absolute minimum of 5,040 that was tentatively established earlier. The ratio of 40,320 to 5,282 is 7.6, very close to 8, a result to be expected based on the same arguments offered for the bishop problem.

Further results were obtained by Smith for larger boards. For instance, a maximum of 9 rooks can be placed on a 9×9 board so as to challenge or occupy every square and yet not attack each other. There are 9!, or 362,880, ways of arranging these 9 rooks of which 46,066 are distinct solutions. Smith's work with the 8-rook problem has been verified independently in Los Angeles, California, by Donald B. Charnley.

This study of chessboard placement problems is far from exhaustive. Much remains to be learned, and, if a suitable symbolism can be devised, many of the unsolved problems will eventually yield their solutions. Some of the arrangements are relatively simple, though proofs of uniqueness or completeness are usually too complex for the novice. Much too little work has been done on chessboard-placement problems, and use of electronic computers could very well produce some interesting results.

3

fun with paper

This chapter will provide a change of pace. Again, a form of geometry is involved, but this time shape rather than placement is the keynote. No elaborate equipment is needed to perform these experiments with paper. The list includes several sheets of paper (scrap paper will do for much of the work), scissors, paste or Scotch tape, and a pencil or pen. Many of the recreations have to do with the effects gained by folding, pasting, and cutting paper, but because there is more interest here in the properties of the configurations formed than in the artistic shapes that can be made, *Origami,* the Japanese art of folding paper to form specific objects, will not be dealt with.

GEOMETRIC CONSTRUCTIONS WITH PAPER ONLY

Paper, used by itself, can reproduce most of the constructions in geometry without the use of ruler, compass, or pencil. Paper can be folded so that the crease forms a straight line or passes through one or two points. Points and lines can be superimposed by folding, and equal lines or angles can be made by duplicating folds. As is the case with ruler, compass, and pencil, these assumptions are true only theoretically. In actuality, the work may not be accurate since our physical ability to fold perfectly is limited, just as is our ability to draw a true straight line or circle no matter how well-constructed the instruments are. Circles and other curved figures cannot be constructed in this way, but curves can be approximated by

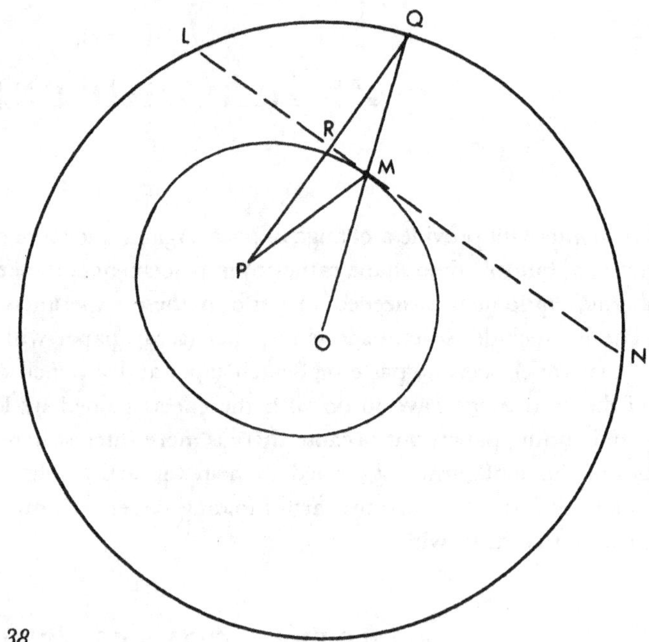

FIGURE *38*

proper folding that produces a pattern of straight-line creases that appears to be curved. Precut pieces of paper must be used, however.

Cut out a circular piece of paper about six inches in diameter and mark a point, P, on the interior (see Figure 38). Fold the paper so that the circumference of the circular disc of paper falls on P. Unfold and repeat the process several times so that different points of the circumference fall on P. After various points are utilized, the crease formed will enclose an ellipse. This can be proved geometrically by referring to the drawing. OQ is any radius, and LN is the fold that brings Q into juxtaposition with P. Right triangles QRM and PRM are congruent and therefore $MQ = MP$. Then, $OM + MP = OM + MQ = OQ$, and OQ will be a constant for all positions of Q. This means that M lies on an ellipse whose foci are O and P. Since angles PMR, QMR, and OMN are equal, LN is indeed *tangent,* that is, it touches the ellipse at one point.

The *foci* of an ellipse are two points within an ellipse such that any line from one of these points and reflected from the ellipse will pass through the other point. If a pool table were made in the form of an ellipse, a ball starting from one focus would pass through the other one after a single bounce from the side of the table. Actually, such a pool table is available. It is called Elliptipool and is manufactured by the Gotham Educational Equipment Co., New Rochelle, New York. There is only one pocket at one of the foci, and the strategy of a game of elliptipool is based on the reflection properties of the ellipse.

A parabola is a familiar, horseshoe-shaped curve. A ball thrown through the air traces a nearly perfect parabolic curve as it falls back to earth. The cross section of the reflecting surface of an automobile headlight is parabolic. A light placed at a certain point inside a parabolic reflecting surface will send out reflected light rays that are nearly parallel. This certain point is called the *focal point* of the parabola.

A parabola can be formed from a piece of paper with a straight edge. Fold the paper repeatedly so that various points on one straight edge coincide with a fixed point, P, on the paper. The creases will envelope a parabola, with the focal point, or focus, at P and the directrix on the edge of the paper that was used during the folding process. The *directrix* of a parabola is a straight line such that the length of a line perpendicular to it and cutting the parabola at a point, say M, is the same as the distance from point M to the focal point of the parabola. That such a parabola results by the folding process described can be proven by referring to Fig-

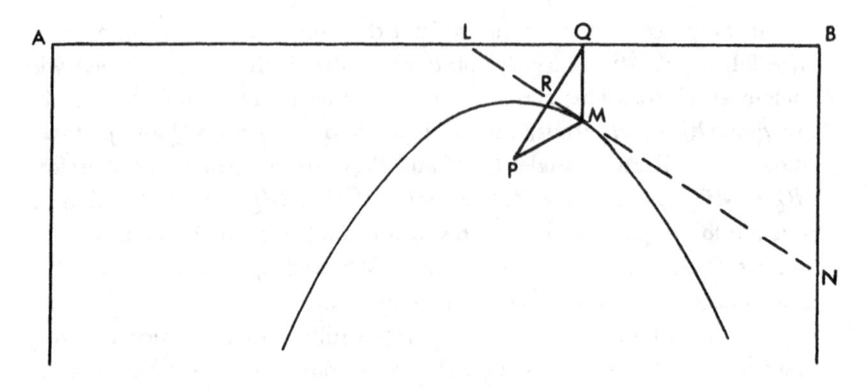

FIGURE 39

ure 39. *LN* is the fold that brings *Q*, any point on the straight edge *AB*, into coincidence with *P*. A perpendicular from *Q* intersects *LN* at *M*. Right triangles *QRM* and *PRM* are congruent, and therefore $QM = PM$. Then *LN* is tangent to the parabola at *M*.

In any ellipse, the sum of the distances from any point on the ellipse to both the foci is a constant. In any parabola, the distance between the focus of the parabola and any point on the curve is equal to the distance from that point to the directrix. Since these conditions are fulfilled by the described methods of folding, the curves are approximated by the patterns of straight-line creases resulting from the paper-folding processes.

Regular polygons can also be constructed by folding strips of paper into knots. Figure 40 shows how an equilateral triangle may be formed

FIGURE 40

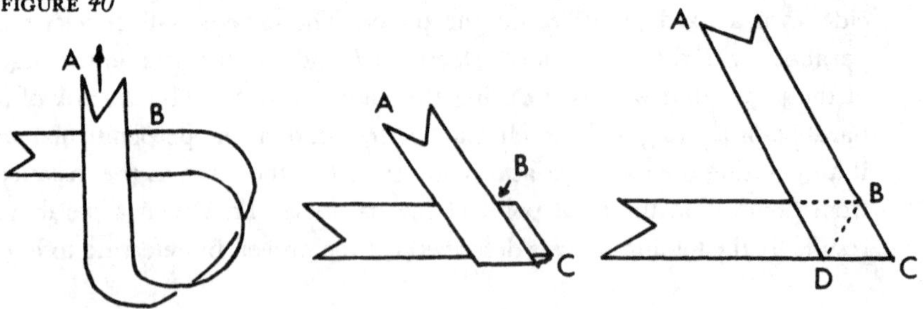

from a strip of paper. Place the edge of the end, *A*, on some point, *B*, and slide this edge upward, always passing along point *B*, until a funnel-shaped form results with its peak at *C*. If this funnel is creased along *BC* and the strip turned over, the equilateral triangle *BCD* is seen.

A square can be formed by folding a strip of paper on itself and making a crease (see Figure 41). Open the strip and fold one end down along this crease and over. A square will result if end *A* is folded across and back as shown in the figure.

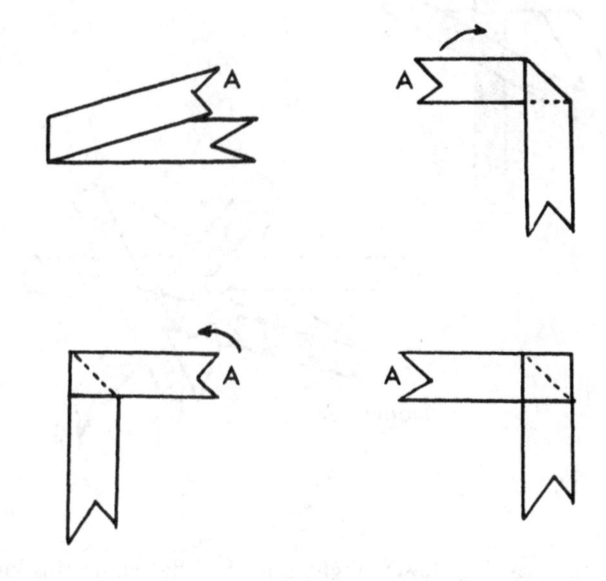

FIGURE *41*

More complicated figures can be formed by making slightly more intricate folds. If a strip of paper is simply knotted, as shown in Figure 42*a*, a regular pentagon will result upon the tightening and flattening of the knot. A regular heptagon can be formed by starting with the same simple knot and bending end *B* away from you, up and under the whole configuration, coming out near *A*. Then fold *B* over all the white portions, as

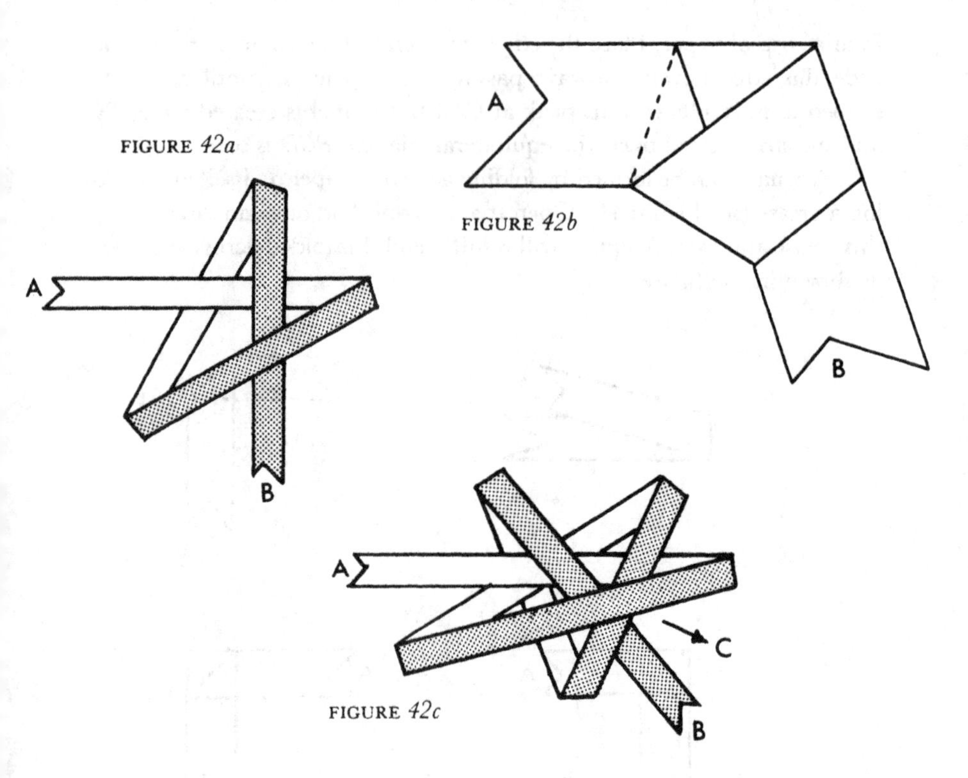

FIGURE *42a*

FIGURE *42b*

FIGURE *42c*

42*c* in the drawing shows. Tightening and flattening this knot produces the heptagon shown in 42*d*.

This knotting process can be continued to produce a regular nonagon, a nine-sided polygon. Referring again to Figure 42*d*, bend the end *B* of the strip away from you and bring it up and under the whole configuration to a point near *A*. Now feed end *B* over all the white portions and under all the shaded sections, coming out at *C*, the point shown in Figure 42*d*. This loose knot can be tightened and flattened into the regular nonagon of Figure 42*e*.

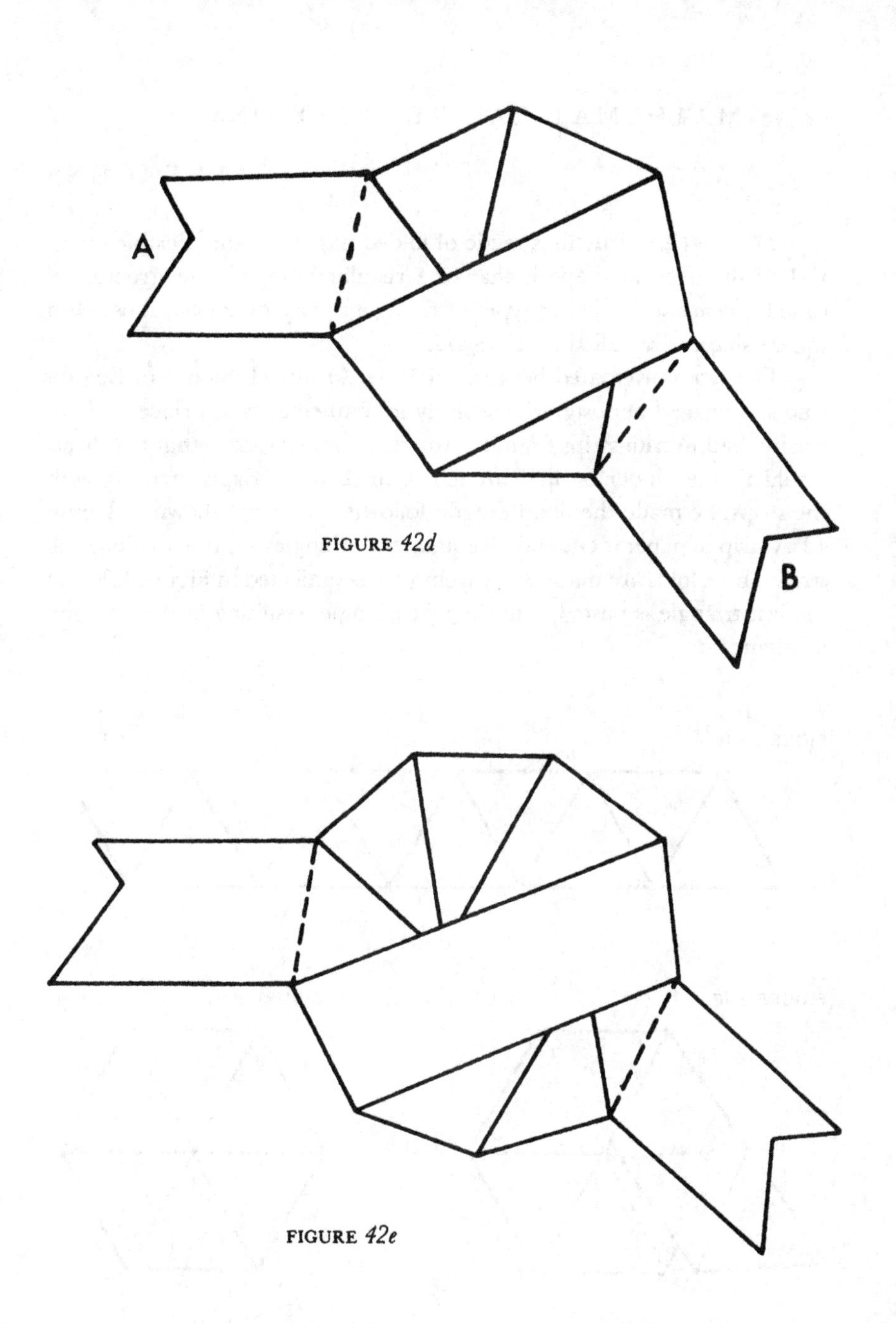

FIGURE *42d*

FIGURE *42e*

FLEXAGONS

Flexagons are structures made of folded paper or other flexible material. If the over-all shape is that of a regular hexagon, the structure is called a *hexaflexagon.* Other types of flexagons exist, of course; those with square shapes are called *tetraflexagons.*

Flexagons originated because, in 1939, Arthur H. Stone, an English student engaged in postgraduate study in mathematics at Princeton University, had to trim strips from his American note paper so that the sheets would fit the binder he had brought from Britain. Experimenting with the strips, he made the first flexagon following the steps shown in Figure 43. A strip of paper is cut and 10 equilateral triangles are drawn along the strip. Three folds are made away from you as indicated in Figure 43*b*, and the last triangle is pasted onto the first triangle, resulting in the flexagon in Figure 43*c*.

FIGURE 43*a*

FIGURE 43*b*

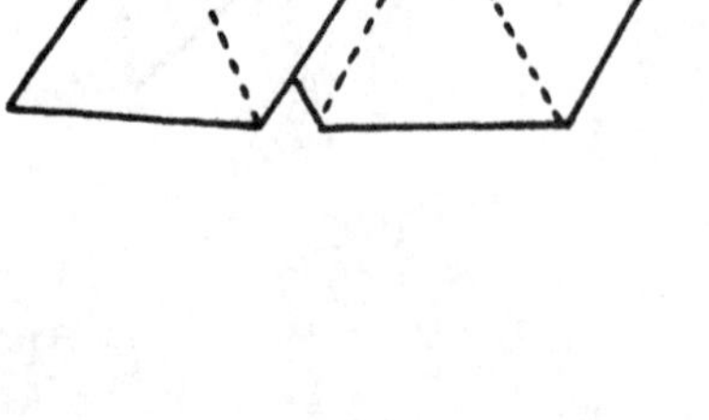

FIGURE 43*c*

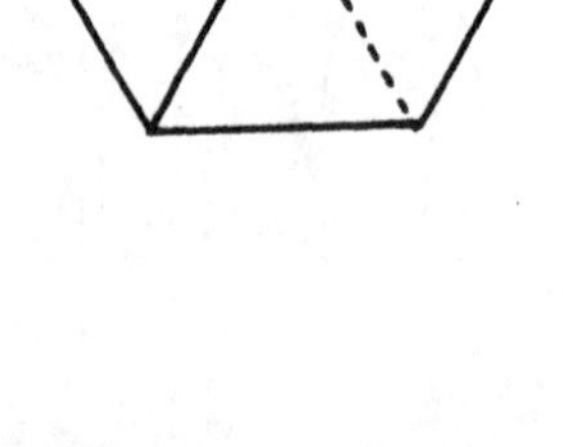

FIGURE *44a*

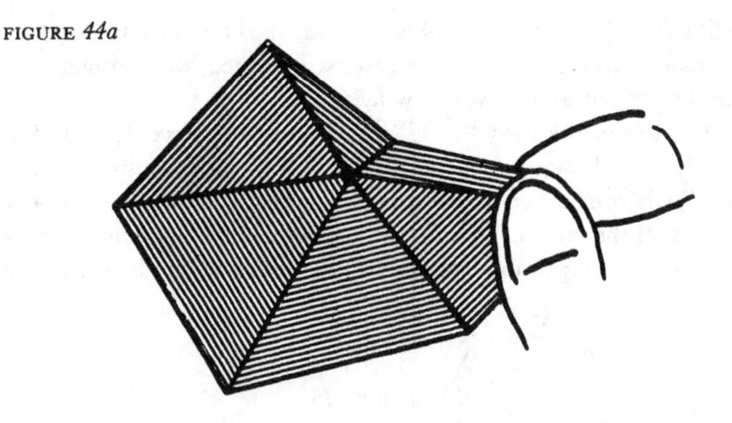

A flexagon has the property that it may be folded into a shape like that of the wings of a dart and then reopened at the center revealing an entirely new face at the expense of hiding the original one. In the drawings of Figure 44, the original face is indicated by the straight-line patterns, the new face, about to appear, is indicated by the dotted surface. A flexagon may be opened or flexed by taking hold of two triangles and pinching them together, as shown. An edge of the opposite set of triangles

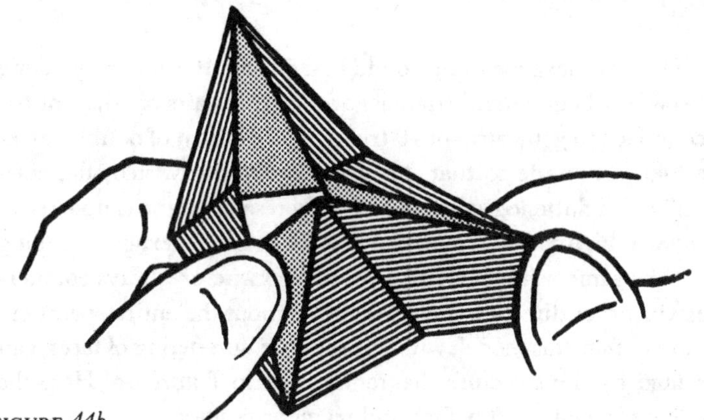

FIGURE *44b*

can be lifted up. If flexing does not now occur, start by pinching together another pair of triangles. Once the process is started, the flexagon will open inside and out and show a new face.

If the faces of the model in Figure 43*a* are numbered 1, 2, and 3, respectively, the flexagon may be represented by a structure diagram (Figure 45) where each vertex corresponds to a face and each line segment to a position of the model. The *structure diagram* is merely a planar representation of the flexagon that includes all the information about its folds

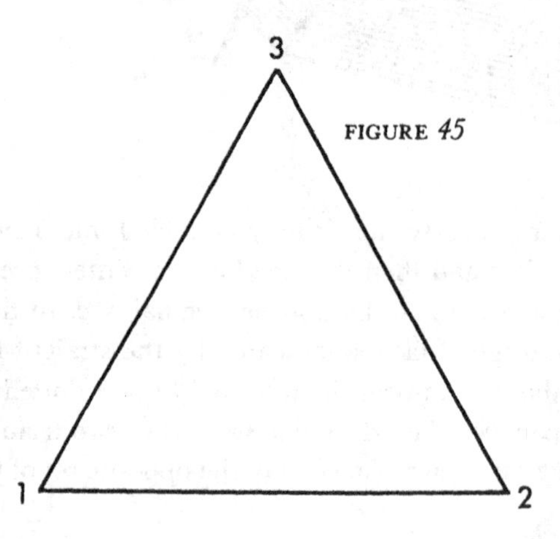

FIGURE *45*

and faces. The next flexagon Stone found had 6 faces. It is made by taking a straight row of 19 equilateral triangles and folding pairs of adjacent triangles face-to-face to get a strip of 10 triangles, 9 of them of double thickness. The folds are made so that the strip becomes a twisted, flattened spiral. Finally, this flattened spiral (which now resembles the simpler strip of 10 triangles in Figure 43*a*) is folded as was the original flexagon, keeping the twist in the same sense (that is, always clockwise or always counterclockwise, whichever direction is chosen) throughout the entire operation. It will be found that this model, with appropriate numbering of faces, can be represented by the structure diagram shown in Figure 46. Here the correspondence is again vertex-face and segment-position.

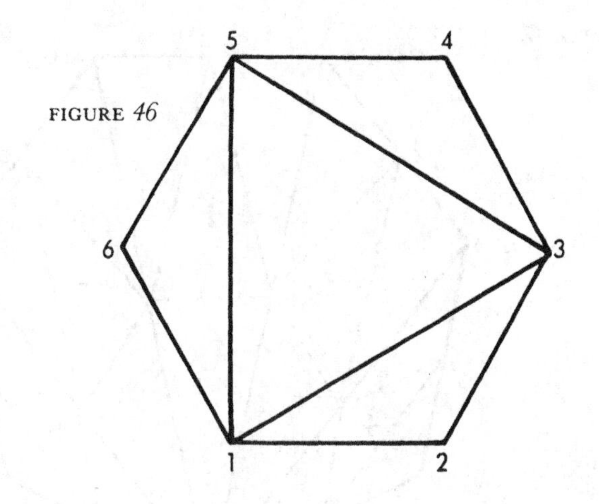

FIGURE 46

Figure 46 is obtained from Figure 45 by applying one triangle to each side of the original triangle. In general, the figure obtained by applying a triangle to one side of a given triangle, then applying a second one to an outside segment of the resulting figure, and repeating this operation as often as desired, is a polygon subdivided into triangles by nonintersecting diagonals. This raises the question: Can a flexagon be made that corresponds to any such structure diagram? It has been answered in the affirmative by Stone and others.

The originators of flexagons developed a method of construction such that the precise instructions for making a model could be derived from the structure diagram, and the model, in the terms given above, exhibits only those properties implied by the diagram. While Stone's original method is superior if extensive treatment and rigorous proofs are desired, the instructions that follow are simplified for actual construction. They were developed by a puzzlist, Sidney H. Scott of Watford, England. Though these instructions are "simplified," they are not easy; close attention must be paid to the discussion and great care taken in making the models. The beginner in the field may wish to skip this section now, and

FIGURE 47a

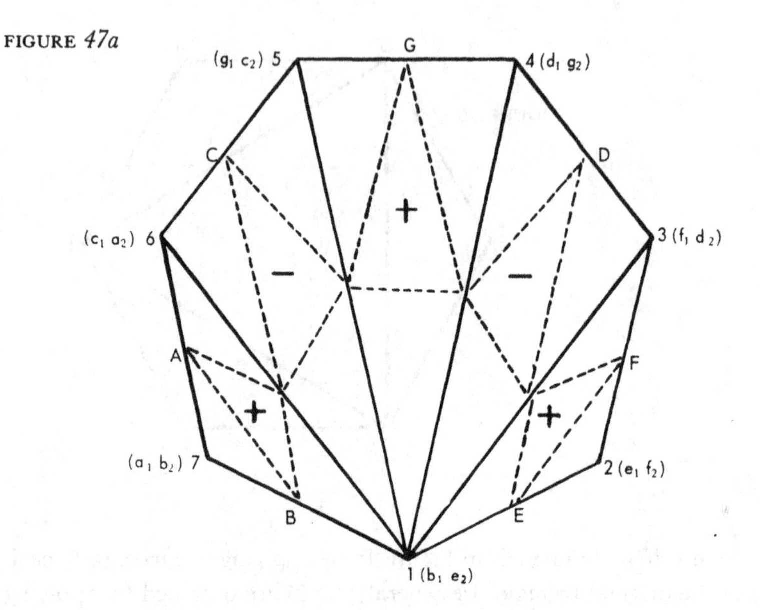

indulge in the other flexagon recreations given on pages 76–81. Familiarity with actual flexagons may make it easier for him to follow this section when he returns to it.

The general case of order n shall be dealt with first. This corresponds to a structure diagram in the form of an n-gon subdivided into $n - 2$ triangles by $n - 3$ nonintersecting diagonals. For purpose of illustration, the fully worked-out plans for one case each of orders 7 and 8 are given—there being a small difference in the routines for even-n and odd-n numbers of triangles. Note that, since to each possible structure diagram there corresponds a flexagon—unique, too, unless right- or left-handedness is regarded as a significant distinction—there are several possibilities for each n if n is equal to or greater than 6. For example, if n is 9, there are 27 different possible flexagons.

Start by drawing a structure diagram with n vertexes. Number the vertexes 1, 2, 3, . . . n in any order. Choose any point, about in the middle perhaps, on each of the $2n - 3$ line segments. These line segments are those that are equal to the total number of the sides of the polygon and the interior nonintersecting diagonals. With these points as vertexes, draw another group of $n - 2$ triangles, so that each triangle of the structure diagram itself contains just one of the new triangles. In Figures 47*a* and *b*, the new triangles are shown in dotted lines. The sides of the triangles in this system constitute a path that crosses itself on each diagonal. It is, of course, possible to define the path so that it does not intersect itself on a particular diagonal, but it is necessary that this interpretation be rejected at each junction for purposes of clarity.

Every possible structure diagram must contain at least 2 triangles, each having 2 sides that are also sides of the polygon. Choose one of these triangles and letter one of its nondiagonal sides *A* and the other *B*. This

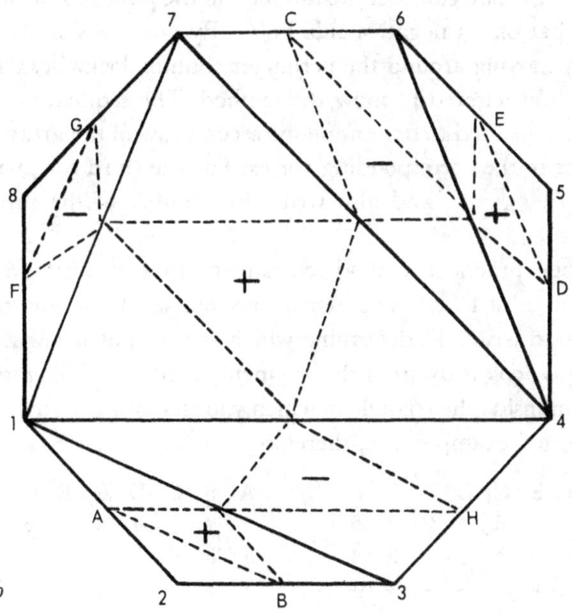

FIGURE 47*b*

must be done so that, if it is imagined that a point moves from A to B along the dotted path by the direct route, the inner triangles are being described in the positive sense (counterclockwise). Now, suppose that the moving point continues along this closed path, crossing itself on each diagonal, until it arrives back at A, having described each inner triangle just once. Within each of these triangles write a plus or a minus sign according to the sense (counterclockwise or clockwise) in which the moving point described it. Letter the other sides of the polygon C, D, E, . . . (n letters, altogether) in alphabetical order corresponding to the order in which they are reached by the moving point. The resulting structures (Figures 47*a* and *b*) will be referred to as the *augmented structure diagram.*

The next step consists of writing out a $4 \times n$ array that will contain all the information necessary to design the strip to be folded into a flexagon. For the first row, write down the n letters A, B, C, ... Each letter of the augmented structure diagram occurs between two numbers on the perimeter of the polygon. Let the number on the positive side of any letter, P, be p_1, that on its negative side be p_2. By positive side is meant that reached by moving around the perimeter counterclockwise. Figure 47*a* is numbered and lettered to show the method. The sequence a_1, b_2, c_1, d_2, e_1, f_2, g_1, . . . , is used to determine the second row of the array by writing the number of the corresponding vertex. For the third row, write: a_2, b_1, c_2, d_1, e_2, f_1, g_2, . . . , and also write the number of the corresponding vertex.

The best practical method seems to be that of putting in all those with the subscript 1 first. The fourth row consists of plus and minus signs and is easy to write. To determine which sign to put below a particular letter, simply look it up from the augmented structure diagram: you will find the sign inside the triangle of which your letter is a vertex. The arrays for the selected examples are, therefore:

A	B	C	D	E	F	G
7	7	6	3	2	2	5
6	1	5	4	1	3	4
+	+	−	−	+	+	+

A	B	C	D	E	F	G	H
2	2	7	4	6	8	8	3
1	3	6	5	5	1	7	4
+	+	−	+	+	−	−	−

Because of a restriction on choice made previously (namely, A and B are vertexes of the same triangle), the fourth row of the array must begin with at least two plus signs. If there are n plus signs, the next stage can be omitted, for the strip will be straight.

Work on the actual model can be started at this time. Take a large sheet of fairly stiff paper and cover one side of it with a network of accu-

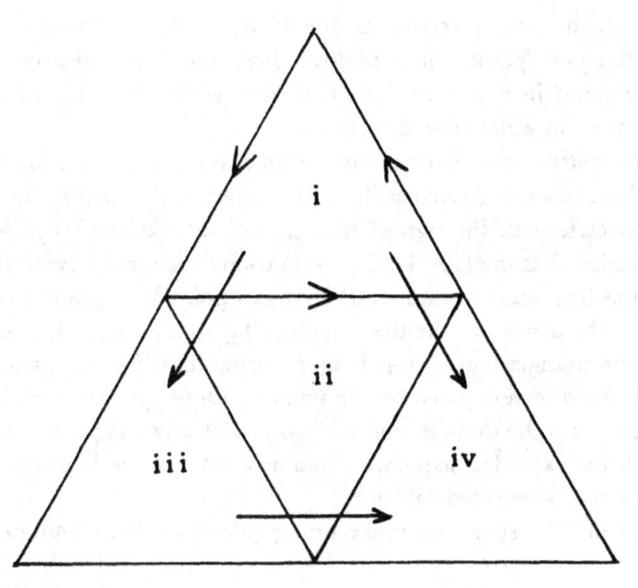

FIGURE *48*

rately drawn equilateral triangles. The area of paper required increases with n, not only because $3n + 1$ triangles are required, but also because, if the model is to move satisfactorily, the minimum possible size of triangle is also an increasing function of n. A convenient size is a triangle of about 1½ inches on a side. It is often more convenient to get an idea of the shape of the strip by using scratch paper first, and here it is quite easy to work without drawing an actual network. Start with one equilateral triangle and put an arrow on each of its sides, all pointing around the triangle in the same direction. Call these directions positive and those directions exactly opposite will be called negative. The idea is to choose $3n + 1$ triangles that make up a strip in such a way that moves made along the strip from one triangle to its neighbor are made in a sequence of directions given by the fourth row of the array. Two adjacent triangles do not define a direction, but three do. In Figure 48, the order of triangles, *i*, *ii*, *iii* "point" in the same direction as one of the arrows on triangle *i*, and so these triangles "point" in a positive direction. The order of triangles *iii*, *ii*, *iv* point in a positive direction also, while the order of triangles *i*, *ii*, *iv* point in a negative direction.

The fourth row of the array, taken three times over, produces an ordered sequence of $3n$ signs. The first triangle of the strip is the arrowed one. Associate with the sign of triangle i of the extended sign sequence the direction that the $(i - 1)$, i, $(i + 1)$ triangles must follow in the strip. Thus, the first, second, and third triangles point in a positive direction based on the arrows in the first triangle. Then, for $i = 3, 4, 5, 6, \ldots 3n$ obtain the triangle that is $(i + 1)$ by ensuring that $(i - 1)$, i, and $(i + 1)$ point in the direction given by the sign of i. Once the first 3 triangles are chosen, this can be done in one way only. The advantage of this method is that if the extended sequence contains k consecutive like signs, then k triangles can be selected together.

Cut out the strip and make creases along all the common sides of

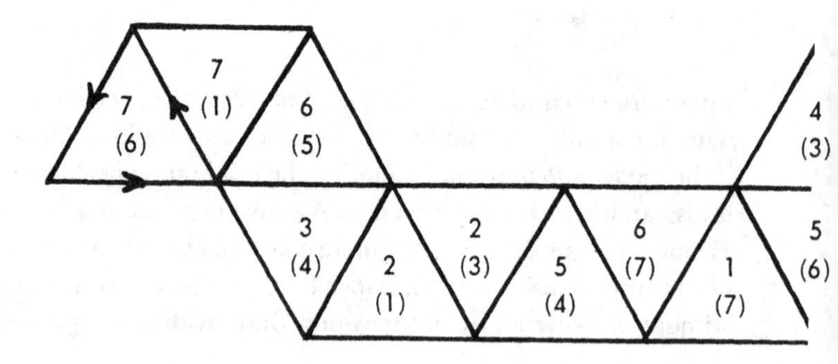

FIGURE 49

adjacent triangles. The shapes for the examples are shown in Figures 49 and 50. The second number, in parentheses, on each triangle is meant to be written on the back, and the numbering is arrived at in one of the two following ways, depending on whether *n* is even or odd:

I. *n* even: Write the numbers of the second row of the array in order, one on each successive triangle on one side of the strip, starting with the first one. Repeat the array twice so that the first $3n$ triangles have been numbered. Write the number a_1 (which is the same as b_2) on the last triangle. In like manner, use the numbers of the third row to label the other side of the triangles, being careful to start again at the first one.

II. *n* odd: The ordered sequence of numbers to be put on one side of the strip is made up of the second row of the array, then the third row, the second row again, and then a_2. The sequence for the other side is made up of the third row of the array, the second row, the third row again, and then a_1. The strip should now bear each symbol 6 times, except for the 2 triangles that appear on the end. These symbols appear 7 times each. When you have completed the flexagon, these end triangles will combine, and the seventh appearance of each of the end symbols will be eliminated.

All that remains to be done now is to fold and seal the model. It will flex more easily if a narrow ribbon around the perimeter is trimmed off. As implied previously, there must be at least 2 vertexes of the structure-

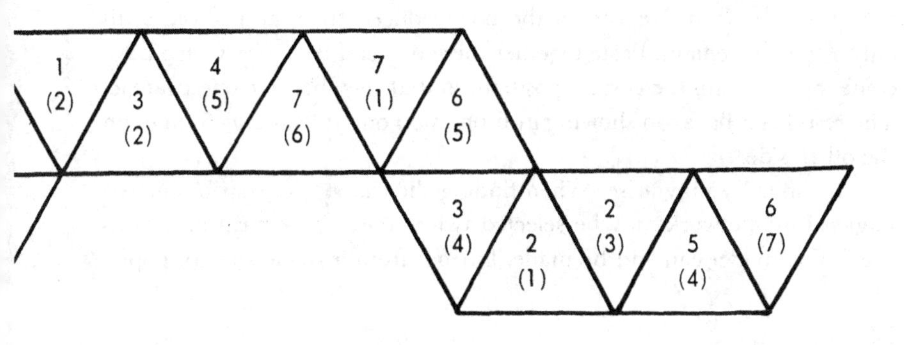

FIGURE 50

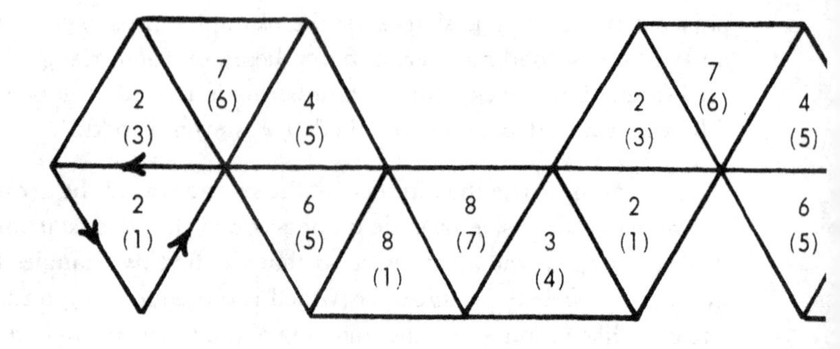

diagram polygon that are not met by a diagonal. One of them is numbered a_1 ($a_1 = 7$ in Figures 47*a* and *b*).

Suppose that one of the other such vertexes bears a number c ($c = e_1 = 2$ in Figure 47*a*). The c occurs on 3 pairs of adjacent triangles of the strip. Fold the triangles of each pair face to face so that c is hidden everywhere. *If* these faces were pasted together (but do not do it), a strip for a flexagon of order $n - 1$ would be produced. Moreover, the diagram for the smaller model is obtained by removing the triangle with vertex c from the diagram of order n. This is a hint toward Stone's inductive proof of his process.

The rule for folding is: Find a symbol that occurs on pairs of adjacent triangles, but does not occur on the first one. Eliminate that symbol by folding together each such pair. Repeat this operation on the reduced strip thus obtained. Repeat on the now reduced strip, and so on, until only a_1 and a_2 remain. Paste together these two end triangles, which have come together into the correct position, so that they become one triangle. The result is a flexagon showing a_1 6 times on one side and a_2 6 times on the other side.

A difficulty may arise. When finding the shape of the strip, one triangle of the network may be selected twice; that is, the strip may cross itself. The model can still be made, but not from a single strip of paper.

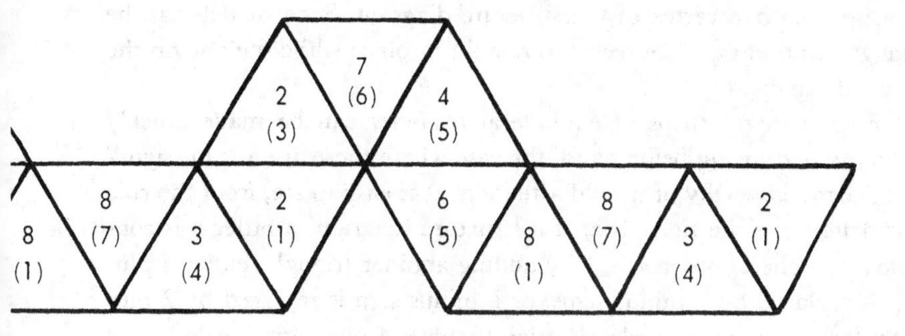

An overlapping one will have to be pasted onto it. The method will be obvious when the problem is met.

Disregarding right- or left-handedness, the correspondence between structure diagrams and flexagons is one to one, but a particular strip may possibly fold into two or more different models. One rather simple example is the seventh-order strip, Figure 49. The reader can renumber the strip and draw the appropriate diagram to give a flexagon that does not correspond to the one shown.

Apart from the mathematical studies that have been made, flexagons have an appeal even for nonmathematicians. These people find fun in trying to arrive at a solution to the number of faces if n is sufficiently large. In order to accomplish this, once a flexagon is finished, the original symbols may be replaced and suitable patterns substituted.

One special type is the street flexagon, whose faces may be so numbered that the symbols 1, 2, 3, . . . n may be made to appear in uninterrupted order. There is one such flexagon for each n, and, in the structure diagram, the diagonals form an unbranched zigzag path. The faces of street flexagons with letters on them instead of numbers can spell out names and be used for unique greeting cards. Each of the unique flexagons of orders 3, 4, and 5 is a street flexagon. Furthermore, they also belong to the same class as the order 7 example of Figure 47*a*, where the diagonals all arise from one vertex of the structure diagram. Such models can be arranged so that one face seems to remain in place while the one on the other side changes.

Since straight strips of equilateral triangles can be made quickly without any drawing beforehand, the case where there are n "plus signs" in the array is worthy of special attention. It is easy to see, from the rule that determines the signs, that if a structure diagram of order n is converted into one of order $n + 1$ by adding another triangle, either a plus sign is replaced by 2 minus signs, or a minus sign is replaced by 2 plus signs. It follows that a single plus sign becomes 4 plus signs if side 1, 5 of the figure 1, 2, 3, 4, 5 (part of Figure 46) is applied to any positive side of a structure diagram. Hence, if this is done to an nth-order structure diagram of a straight flexagon, the diagram for another straight model will have been drawn, this time of the order $n + 3$. A second method of conversion, this time from order n to $2n$, is achieved by building a triangle on every side of the original diagram (compare Figures 45 and 46). The n plus signs become $2n$ minus signs, and this is again a condition for a straight strip. This practical method of making models has been formulated with some loss of generality.

It is easy to see that the order of a straight model is always a multiple of three. The first flexagons ever made were members of a subclass of straight-strip flexagons. Here n is of the form $n = (3)(2^q)$, where q is any nonnegative integer. The strips can be folded quickly by an extension of the short method, that is, by successive windings each of which halves the length of the strip. The structure diagrams for this class are obtained by using the second method of conversion (n to $2n$) as many times as necessary, starting from order 3. However, any structure diagram of a straight flexagon can be developed by using only the first method of conversion (n to $n + 3$). In this way, it can be shown that there is only one straight flexagon each of the orders 3, 6, and 9; while there are 4 of order 12, and 14 of order 15.

The structure diagrams of the 4 straight flexagons for $n = 12$ are shown in Figure 51, from which it may be seen that the diagonals make

FIGURE 51

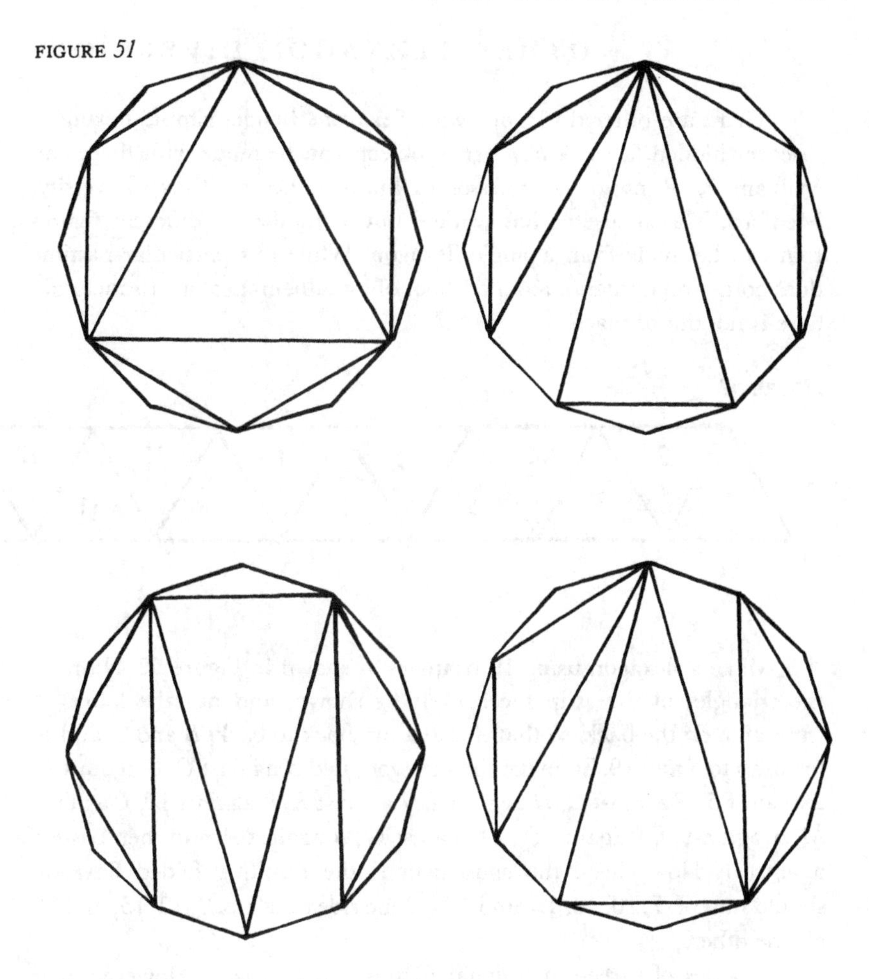

up $\frac{(n-3)}{3}$ discrete triangles, which may be chosen in any way so that they touch vertexes of the polygon and each other only with their own vertexes.

OTHER FLEXAGON DIVERSIONS

There are other diversions with flexagons besides simple flexing to discover hidden faces. A number of objects can be made with them and William R. Ransom, a professor of mathematics at Tufts University, Medford, Massachusetts, has worked out a number of different figures that can be made from a single flexagon. While this particular pastime does border on Origami, as an offshoot of a mathematical art its inclusion here is not out of place.

FIGURE *52*

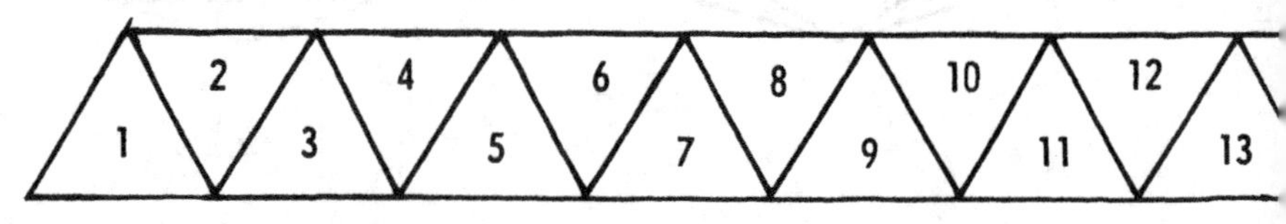

Make a flexagon using 19 triangles as shown in Figure 52. Number the triangles in the strip successively as shown, and put the letters *A* through *S* on the back, so that *A* and 1 are back to back, *B* and 2, and so forth up to *S* and 19. To make the flexagon, fold *B* against *C*, *D* against *E*, 2 against 5, *F* against *G*, *H* against *I*, *J* against *K*, 8 against 11, *L* against *M*, *N* against *O*, *P* against *Q*, 14 against 17, *A* against *R*, and then paste *S* against 1. Now check the construction: the resulting folded flexagon should have 4, 7, 10, 13, 16, and 19 on one side and 3, 6, 9, 12, 15, and 18 on the other.

A series of three-dimensional figures can be made. However, it is very easy to go astray, and the directions must be followed carefully. Some unsought-for shapes may turn out to be more interesting than those given here.

Boat. Fold the flexagon along the diagonal that divides 4 and 16 from 7 and 13, with these sides outside. Bring 3 and 6 together, holding

4–7 in the right hand with the fold at the top. Bring the corner of 13 that touches 7 toward you and bring 17 against 8 and 14 against 11. Now, with the long side up, open the boat, separating *I* from *H* (Figure 53).

FIGURE 53

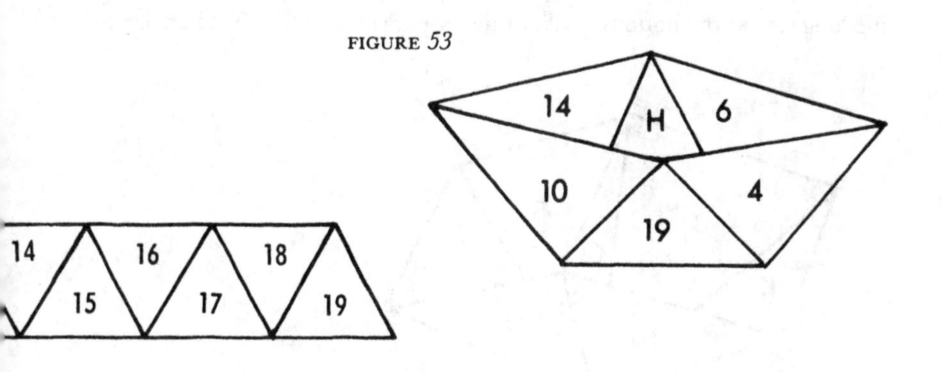

BUOY. Bring the bow and stern of the boat together so that the upper edge of 7 touches the upper edge of 13 and the upper edge of 4 touches the upper edge of 10 (Figure 54).

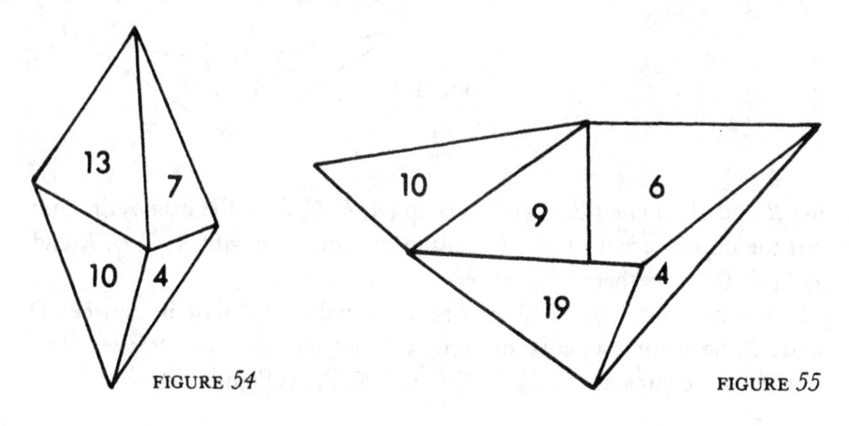

FIGURE 54

FIGURE 55

SQUARE CUP WITH A HANDLE. Return to the boat, and hold 10–13 in the left hand. Push the corner of 4 that touches 10 toward you and down, and you will have a square cup with 3, 6, 9, and 18 on the inside and its handle in your left hand (Figure 55).

TWIN CUPS. Flatten the square cup, bringing 3 against 6 and, with the long side at the bottom, pull out the upper corner of 4–19. Then, keep-

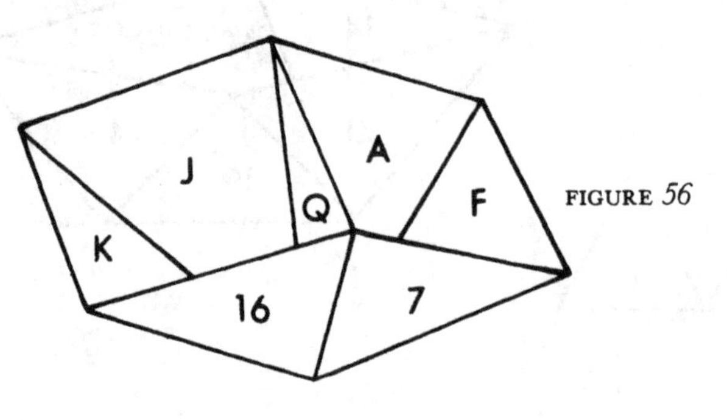

FIGURE 56

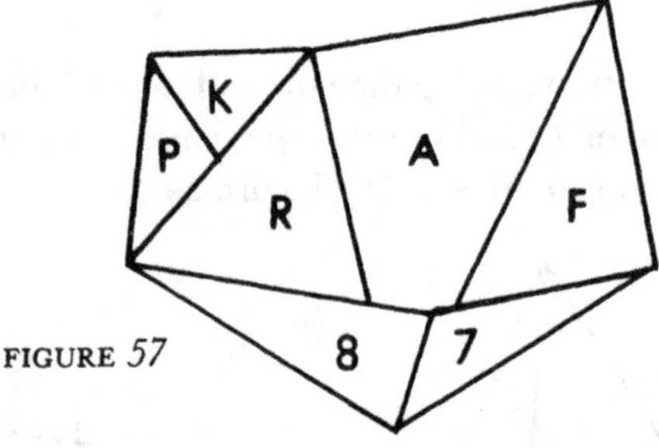

FIGURE 57

ing *R* and 10 in one plane and the cup (*A*, *F*, *G*, *R* on the inside) up, pull out the upper corner of 13–16. This gives twin cups with *A*, *F*, *G*, *R* and *J*, *K*, *P*, *Q* inside them (Figure 56).

PENTAGONAL CUP. Pull on 7 so as to make 10–*J* slide in between *Q* and *R*. The result is a pentagonal cup with a triangular cup attached. The insides of the cups are *A*, *F*, *G*, 9, *R* and *K*, *P*, *Q* (Figure 57).

TRIANGULAR CUPS WITH A HANDLE. Fold *A* against *F*. This makes a handle with 2 cups with *G*, 9, *R* and *P*, *Q*, *K* on their insides (Figure 58).

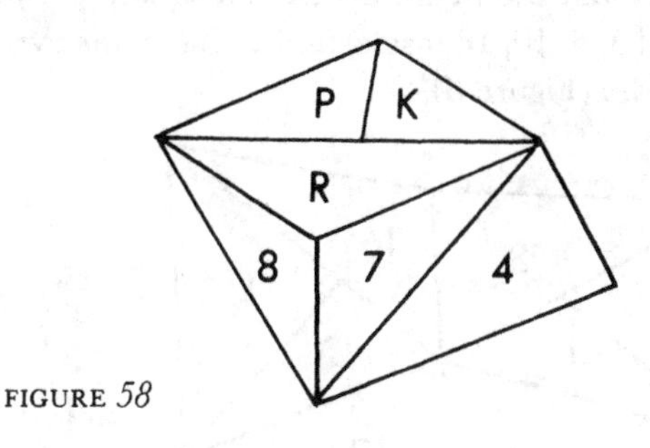

FIGURE *58*

CORNER-TO-CORNER CUPS. Return to the pentagonal cup and fold *A* against *R*. This gives 2 triangular cups, corner to corner, and without a handle (Figure 59).

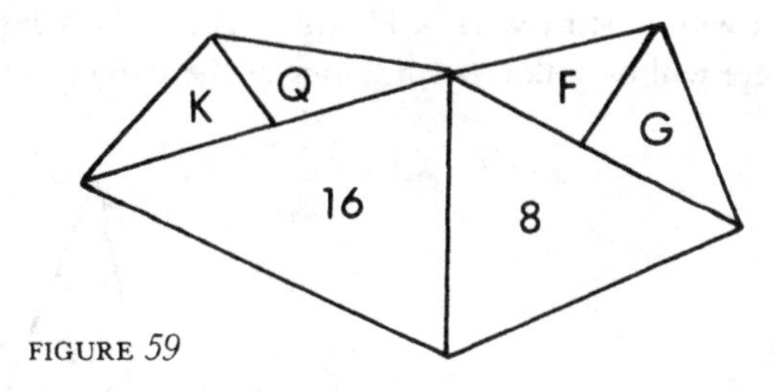

FIGURE *59*

SIDE-BY-SIDE CUPS. Bring 4 against 19. This gives the same 2 cups of Figure 59, but now they will be side by side (Figure 60).

FIGURE *60*

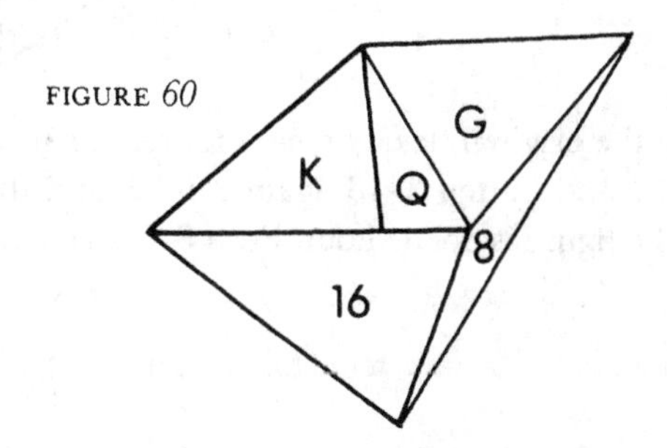

THREE PYRAMIDS. Swing the cups apart until 8 and 16 almost touch, keeping *F* and *Q* in the same plane. Lift the corner of 4–19 at the bottom of the cups, and the figure becomes a square pyramid with 4, *F*, *Q*, 19 outside and 3, 6, 10, 18 inside, and 2 triangular pyramids attached to the *F* and *Q* sides (Figure 61).

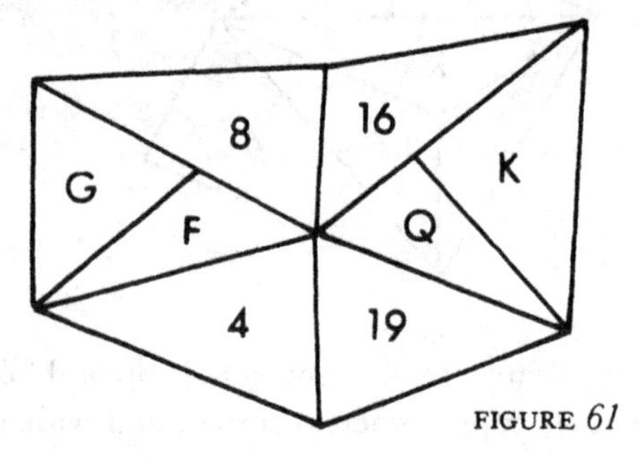

FIGURE *61*

FLOWER BASKET WITH A HANDLE. Bring the sides 3 and 18 together, and there will be a flower container in the two cups (Figure 62).

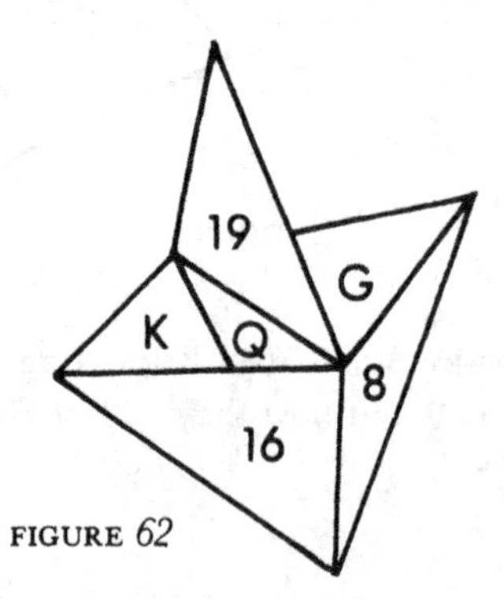

FIGURE *62*

To obtain the original flexagon calls for some painstaking backtracking. Open 4–19 and flatten 3–18 against 6–10, and the pentagonal cup results when the figure is opened out. Push 8–9 through between *Q* and *R* arriving back to the twin cups. Now flatten *A* to *R* and *G* to *F* and flatten *P* to *Q* and *J* to *K*. Hold *F* against *G* and raise the corner of 13 that touches 7; flatten 8 against 11 and 14 against 17. The original flexagon appears when you unfold along the 7–13 diagonal, opening out the edges at 10 and 19.

SOLID FLEXAGONS

This chapter concludes with an introduction to the three-dimensional forms of flexagons developed by Douglas Engel while a student at Kansas State College. He calls these flexagons *flexahedrons*. They are difficult to make, but their unusual flexing properties make it worthwhile to indulge in making a few of the simpler forms. They can be made from tetrahedrons, cubes, and other solids, but only a description of the construction

FIGURE *63*

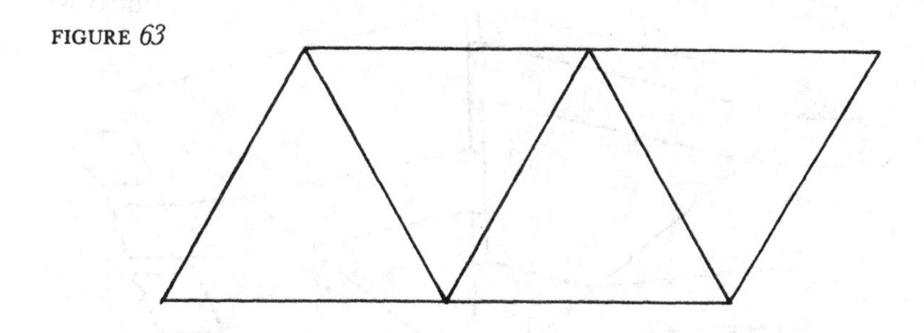

and movements of a closed chain of 6 tetrahedrons, which Engel unmercifully calls a *flexahexatetrahedron*, will be given here.

Use sturdy paper or thin cardboard, and construct 6 regular tetrahedrons by using the form shown in Figure 63, folding into tetrahedrons and binding with Scotch tape as shown in Figure 64. The 6 tetrahedrons are then connected together using strong paper bands as indicated in Figure 65.

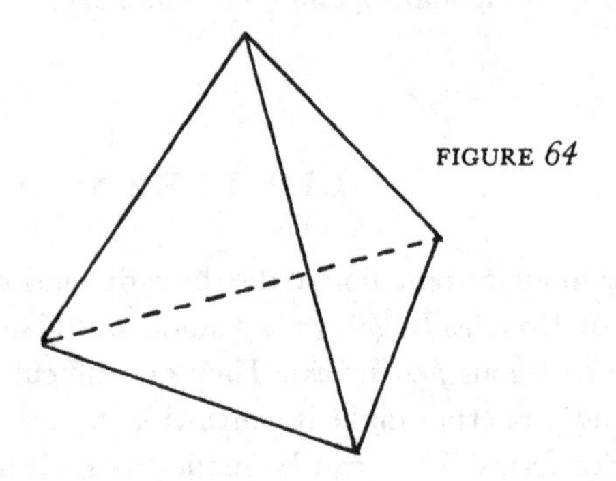

FIGURE *64*

FIGURE *65*

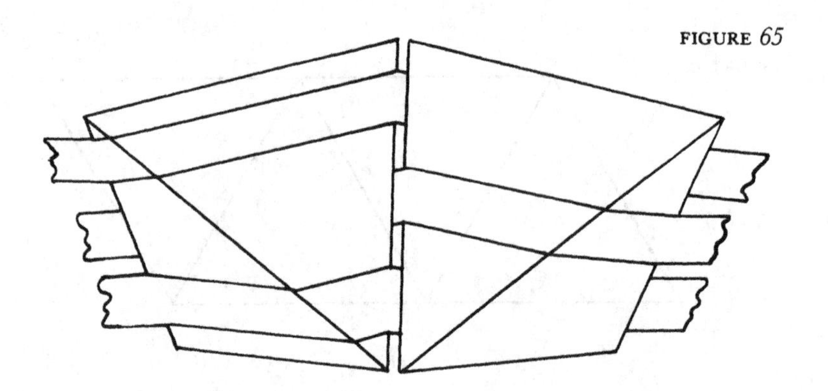

The 6 connected tetrahedrons can be flexed similarly to the flexagons previously described, and Figures 66–69 show the flexing that changes an over-all rectangular shape to a hexagonal one.

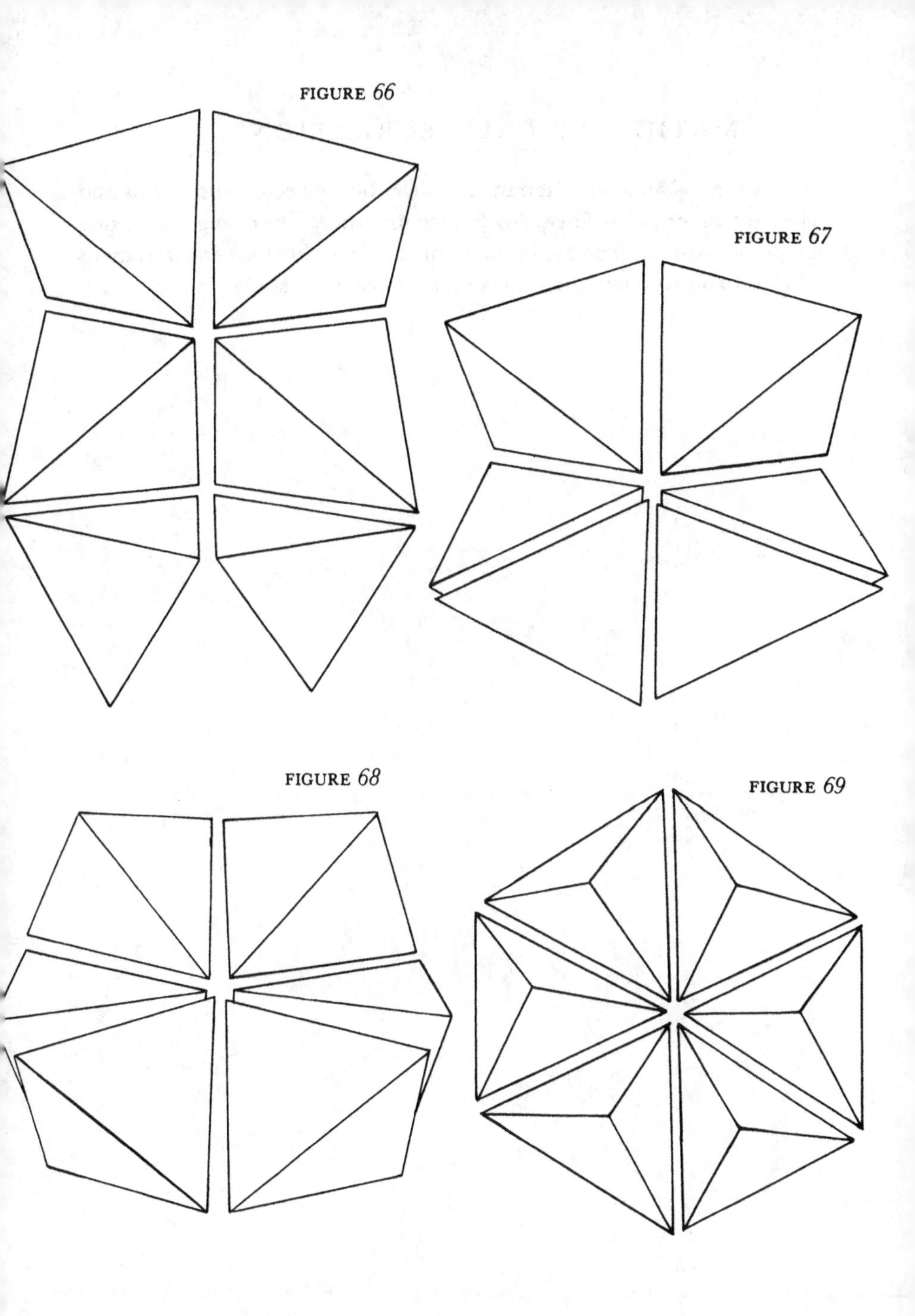

FIGURE *66*

FIGURE *67*

FIGURE *68*

FIGURE *69*

A greater number of tetrahedrons can be connected into chains and given a twist or two before being formed into rings. The changes in flexing properties with an increase in the number of tetrahedrons and variations in chain and ring formation make for a fascinating study.

4

magic and antimagic squares

Magic squares, an ancient and fascinating amusement, are arrays of numbers such that the sum of the numbers in each row, column, or long diagonal is the same. So much has been written about the subject that it would appear that stagnation is about to set in. Still, this entertainment exerts a virtually unbreakable hold on the average recreational mathematician.

MAGIC SQUARES

A magic square of order n is defined as an $n \times n$ array of the integers from 1 to n^2 such that the sum of each row, column, and main diagonal is a constant; that is, all the sums are the same. This *magic constant* is related to n and is equal to

$$\frac{n(n^2 + 1)}{2}$$

If, instead of the consecutive series of integers starting with 1, an arithmetic series starting with, say, 17 and with a difference of 3 between

17	59	56	26
50	32	35	41
38	44	47	29
53	23	20	62

FIGURE *70*

52	61	4	13	20	29	36	45
14	3	62	51	46	35	30	19
53	60	5	12	21	28	37	44
11	6	59	54	43	38	27	22
55	58	7	10	23	26	39	42
9	8	57	56	41	40	25	24
50	63	2	15	18	31	34	47
16	1	64	49	48	33	32	17

FIGURE *71*

successive integers, had been used, a magic square with the terms 17, 20, 23, 26, . . . is formed. The magic constant, C, for any such magic square can be calculated, knowing the order, n, the starting integer, A, and the difference, D, between successive terms:

$$C = n\left(\frac{2A + D(n^2 - 1)}{2}\right)$$

If $A = D = 1$, this equation reduces to $\frac{n(n^2 + 1)}{2}$, which is the magic constant for the usual magic square in which consecutive integers, starting with 1, are used.

Figure 70 shows a fourth-order, that is, 4×4, magic square made up of the series of numbers starting with 17 and with a difference of 3 between successive terms. The equation correctly predicts a magic constant of 158.

Excluding reflections and rotations, there is only one third-order magic square. As *n* increases, however, the number of possible magic squares increases rapidly, as shown in the table below:

n	*Number of Different Magic Squares*
1	1
2	0
3	1
4	880
5	Exact number unknown, but approximately 320,000,000

Magic squares in which the broken diagonals, in addition to the rows and columns and main diagonals, total to the magic constant are called *diabolic* magic squares, or, sometimes, *Nasik* or *panmagic* squares. A broken diagonal in an *n*th-order magic square includes *n* integers along 2 short parallel diagonals. For example, 53–3–4–49–18–40–39–22 or 50–1–4–51–21–38–39–24 in Figure 71 are broken diagonals.

Benjamin Franklin found amusement in magic squares and one of his more remarkable achievements in this area is shown in Figure 71. The magic constant for this square is 260. However, any half-row or half-column totals 130, and the 4 corners plus the middle 4 squares total 260. Franklin's square is also diabolic, since the broken diagonals total 260. There are quite a number of other patterns totaling either 260 or 130. Each of the 4 *bent* diagonals totals 260 (one of these is 52–3–5–54–10–57–63–16). Franklin also composed a 16 $\times$ 16 magic square with equally remarkable properties.

A particularly interesting magic square is shown in Figure 72. The magic constant is 2,056, but the unusual nature of this square lies in its

184	217	170	75	188	219	172	77	228	37	86	21	230	39	88	25
169	74	185	218	171	76	189	220	85	20	229	38	87	24	231	40
216	183	68	167	222	187	78	173	36	227	22	83	42	237	26	89
73	168	215	186	67	174	221	190	19	84	35	238	23	90	41	232
182	213	166	69	178	223	176	79	226	33	82	31	236	43	92	27
165	72	179	214	175	66	191	224	81	18	239	34	91	30	233	44
212	181	70	163	210	177	80	161	48	225	32	95	46	235	28	93
71	164	211	180	65	162	209	192	17	96	47	240	29	94	45	234
202	13	126	61	208	15	128	49	160	241	130	97	148	243	132	103
125	60	203	14	127	64	193	16	129	112	145	242	131	102	149	244
12	201	62	123	2	207	50	113	256	159	98	143	246	147	104	133
59	124	11	204	63	114	1	194	111	144	255	146	101	134	245	150
200	9	122	55	206	3	116	51	158	253	142	99	154	247	136	105
121	58	205	10	115	54	195	4	141	110	155	254	135	100	151	248
8	199	56	119	6	197	52	117	252	157	108	139	250	153	106	137
57	120	7	198	53	118	5	196	109	140	251	156	107	138	249	152

FIGURE *72*

method of construction. If the numbers are followed consecutively, it is found that the "moves" from one to the next are knight moves. That is, if the moves are numbered consecutively and made to traverse each square of a chessboard, or a larger board, only once, then it may be possible to number an $n \times n$ board in such a way as to form a magic square. This would be impossible for any board where n is odd, and it is believed to be impossible for an 8×8 board. The best that has been achieved on the 8×8 chessboard is a semimagic square in which the rows and columns, but not the main diagonals, have magic constants of 260. Certain larger, fully magic squares can be devised with knight moves and Figure 72 is a 16×16 example of just such a square. The tour in this magic square is closed, since the first and last moves are also connected by a knight move; an open-tour magic square is one in which the first and last moves are not so related.

FIGURE *73*

223	283	200	322	163	164
177	408	336	244	12	178
228	122	24	306	448	227
258	36	488	112	204	257
308	224	102	48	366	307
161	282	205	323	162	222

Thousands of magic squares have been constructed and, in a way, if you have seen one, you have seen them all. However, there are a number of not-so-ordinary versions, and some of these are shown on the next few pages.

Figure 73 is a magic square that looks ordinary enough and has a magic constant of 1,355. However, the property that makes this square more than ordinary is the fact that the inner 4 × 4 square is a separate magic square in which the *product* of the integers in each row, column, and main diagonal is equal to 401,393,664 (for example, 408 × 336 × 244 × 12 = 401,393,664). This ingenious magic square, devised by Ronald B. Edwards, an amateur magician of Rochester, New York, exhibits another unusual property. If all the integers in the 4 × 4 multiplication square are reversed (408 becomes 804, 24 becomes 42, and so on) a new multiplication magic square is formed with a magic product of 4,723,906,824.

A square that is completely magic for both addition and multiplication is shown in Figure 74. The addition constant is 1,200, while the multiplication constant is 1,619,541,385,529,760,000. This ninth-order magic

17	171	126	54	230	100	93	264	145
124	66	290	85	57	168	162	23	225
216	115	75	279	198	29	170	76	42
261	186	33	210	68	38	200	135	69
50	270	92	87	248	165	21	153	114
105	51	152	150	27	207	116	62	330
138	25	243	132	58	310	95	63	136
190	84	34	184	125	81	297	174	31
99	232	155	19	189	102	46	250	108

FIGURE *74*

200	87	95	42	99	1	46	108	170
14	44	10	184	81	85	150	261	19
138	243	17	50	116	190	56	33	5
57	125	232	9	7	66	68	230	54
4	70	22	51	115	216	171	25	174
153	23	162	76	250	58	3	35	88
145	152	75	11	6	63	270	34	92
110	2	28	135	136	69	29	114	225
27	102	207	290	38	100	55	8	21

FIGURE *75*

square was constructed in 1962 by Gakuho Abe of Akita-ken, Japan, following a technique given by Walter W. Horner, a retired mathematics teacher of Pittsburgh, Pennsylvania. Horner himself devised a similar ninth-order magic square, but with smaller magic constants. This is shown in Figure 75. The addition constant is 848, and the multiplication constant is 5,804,807,833,440,000.

Figure 76 is a magic square whose constant along any row, column, and main and broken diagonal is 150. Note that the integers from 1 to 36 are not all used, larger numbers being utilized. If any of the patterns shown in Figure 77 are used with Figure 76 the sum of the numbers under the *X*'s is equal to 25 times the number of *X*'s in the pattern. For example,

1	42	29	7	36	35
48	9	20	44	13	16
5	38	33	3	40	31
43	14	15	49	8	21
6	37	34	2	41	30
47	10	19	45	12	17

FIGURE *76*

FIGURE *77*

X	X	X
X	X	X
X	X	X

X	X	X	X
X	X	X	X
X	X	X	X
X	X	X	X

	X	X	
X			X
X			X
	X	X	

X			X
	X	X	
	X	X	
X			X

X	X	X
X		X
X		X
X	X	X

X	X	X	X
X			X
X			X
X	X	X	X

	X		
X	X		
		X	X
		X	

X	X	X	X
X			X
X			X
X			X
X	X	X	X

lay the third pattern (with 6 *X*'s) in the second row of Figure 77 on top of any 4 × 4 set of numbers in Figure 76. One such set would be 48, 9, 42, 15, 33, 3. The sum of these numbers is 150, which is exactly 6 times the number of *X*'s in the pattern.

Two truly unusual magic squares, both composed by a puzzlist who at the time was a prison inmate, are shown in Figures 78 and 79. Both are made up of prime numbers only, which is not too unusual. But the 7 × 7 square has a magic constant of 26,627 along every row, column, and main and broken diagonal. In addition, if the units' places are all removed (that is, 9,341 becomes 934, 6,397 becomes 639, 11 becomes 1, and so on) a new magic square is formed having a magic constant of 2,760 along every row, column, and main and broken diagonal.

11	3851	9257	1747	6481	881	5399
6397	827	5501	71	3779	9221	1831
3881	9281	1759	6361	911	5417	17
839	5381	101	3797	9227	1861	6421
9311	1777	6367	941	5441	29	3761
5387	131	3821	9239	1741	6451	857
1801	6379	821	5471	47	3767	9341

FIGURE *78*

I can only marvel at Figure 79. Here is a 13 × 13 magic square composed of prime numbers only, but the 13 × 13 consists of successive 11 × 11, 9 × 9, 7 × 7, 5 × 5, and 3 × 3 magic squares nested within each other. The magic constants for these individual magic squares are 70,681; 59,807; 48,933; 38,059; 27,185; and 16,311, respectively. The common difference between these magic constants is 10,874, a figure applying even to the difference between the center integer, 5,437, and the magic constant of the 3 × 3 magic square.

1153	8923	1093	9127	1327	9277	1063	9133	9661	1693	991	8887	8353
9967	8161	3253	2857	6823	2143	4447	8821	8713	8317	3001	3271	907
1831	8167	4093	7561	3631	3457	7573	3907	7411	3967	7333	2707	9043
9907	7687	7237	6367	4597	4723	6577	4513	4831	6451	3637	3187	967
1723	7753	2347	4603	5527	4993	5641	6073	4951	6271	8527	3121	9151
9421	2293	6763	4663	4657	9007	1861	5443	6217	6211	4111	8581	1453
2011	2683	6871	6547	5227	1873	5437	9001	5647	4327	4003	8191	8863
9403	8761	3877	4783	5851	5431	9013	1867	5023	6091	6997	2113	1471
1531	2137	7177	6673	5923	5881	5233	4801	5347	4201	3697	8737	9343
9643	2251	7027	4423	6277	6151	4297	6361	6043	4507	3847	8623	1231
1783	2311	3541	3313	7243	7417	3301	6967	3463	6907	6781	8563	9091
9787	7603	7621	8017	4051	8731	6427	2053	2161	2557	7873	2713	1087
2521	1951	9781	1747	9547	1597	9811	1741	1213	9181	9883	1987	9721

FIGURE *79*

Studies of the properties of magic squares inevitably leads to minimum-sum constructions. This is best explained by noting the third-order magic squares in Figure 80 as tabulated by Rudolph Ondrejka, a mathematician from Atlantic City, New Jersey. The characteristics of each of the squares in Figure 80 are given below:

A. This square is composed of primes only and has the smallest possible magic constant, 177.

B. This square is the one with the smallest magic constant (3,117)

FIGURE *80*

71	89	17
5	59	113
101	29	47

A

1669	199	1249
619	1039	1459
829	1879	409

B

18	4	14
8	12	16
10	20	6

C

121	114	119
116	118	120
117	122	115

D

22	4	16
8	14	20
12	24	6

E

27	6	21
12	18	24
15	30	9

F

21	8	16
10	15	20
14	22	9

G

51	9	39
21	33	45
27	57	15

H

63	9	45
21	39	57
33	69	15

I

that is composed of primes in arithmetical sequence. Dudeney first noted this square.

C. This square is composed of composite integers, that is, integers that are not prime numbers, in arithmetic sequence producing the smallest possible magic constant, 36.

D. Dudeney also noted this particular square, which is composed of composite integers in consecutive order and produces the smallest possible constant, 354.

E. This square is composed of composite integers not in arithmetical sequence and has the smallest possible magic constant, 42.

F. This square uses both odd and even composite integers in arithmetical sequence and has the smallest possible magic constant, 54.

G. This square, like *F*, uses both odd and even composite integers, but they are *not* in arithmetical sequence, and the magic constant here is 45.

H. This square uses odd composite integers in arithmetical sequence and has the smallest possible magic constant, 99.

I. This square, like *H*, uses only odd composite integers but they are *not* in arithmetical sequence, and the magic constant is 117.

Figure 80*A* does not include the integer 1, which is not usually considered a prime number. However, if 1 is allowed as a prime, then a magic square composed of primes only can be constructed with a constant of 111.

The reader can try to find similar results for magic squares of higher order. A 4 × 4 magic square similar to that shown in Figure 80*B*, using 16 primes in arithmetical sequence, has been made possible (see page 154).

MISCELLANEOUS MAGIC CONFIGURATIONS

Magic properties of numbers need not be confined to the geometry of the square. Instead of squares, intersecting circles can be constructed so that their points of intersection are numbered magically. Figures 81–83

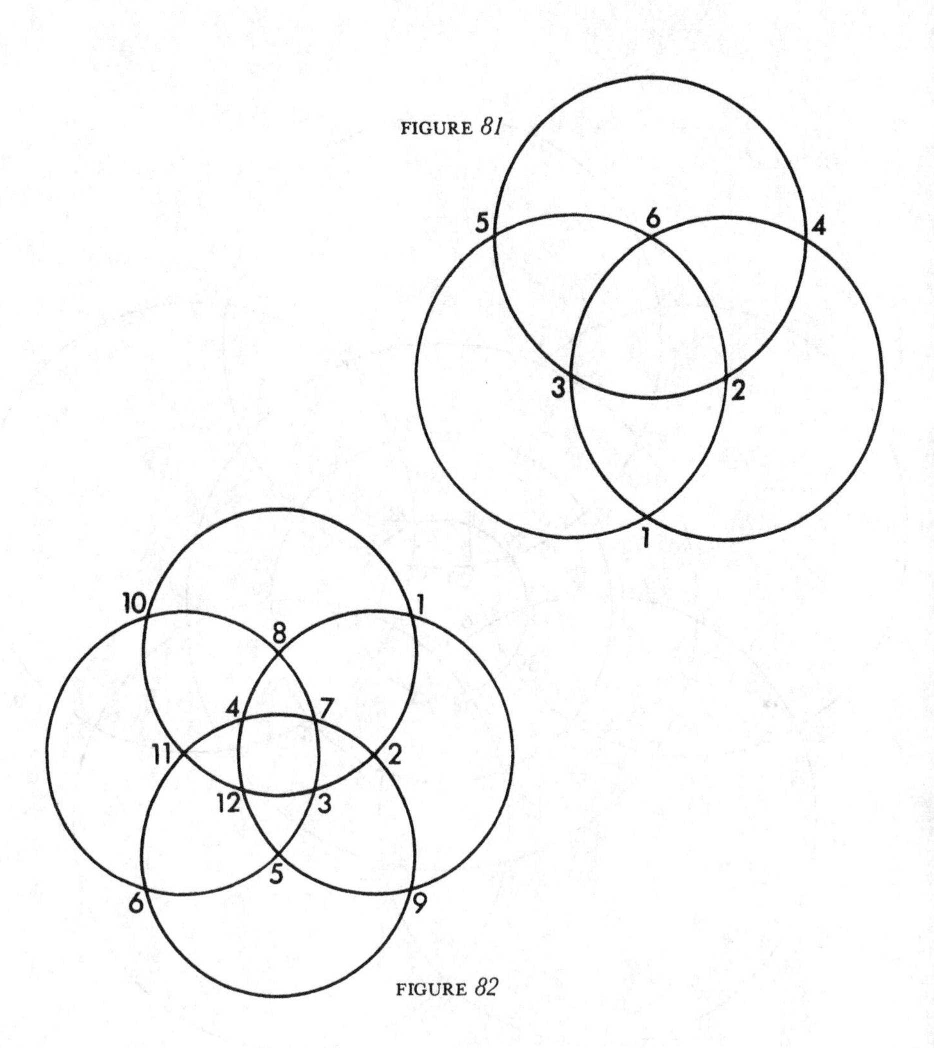

FIGURE 81

FIGURE 82

show 3 magic circles; in each, the sum of the integers lying on any given circle is equal to the sum of the integers lying on any other one. In Figure 82, the sum of the numbers 1, 8, 4, 12, 5, and 9, all on one circle, is equal to the sum of the numbers 6, 5, 3, 7, 8, and 10, all on another.

FIGURE 83

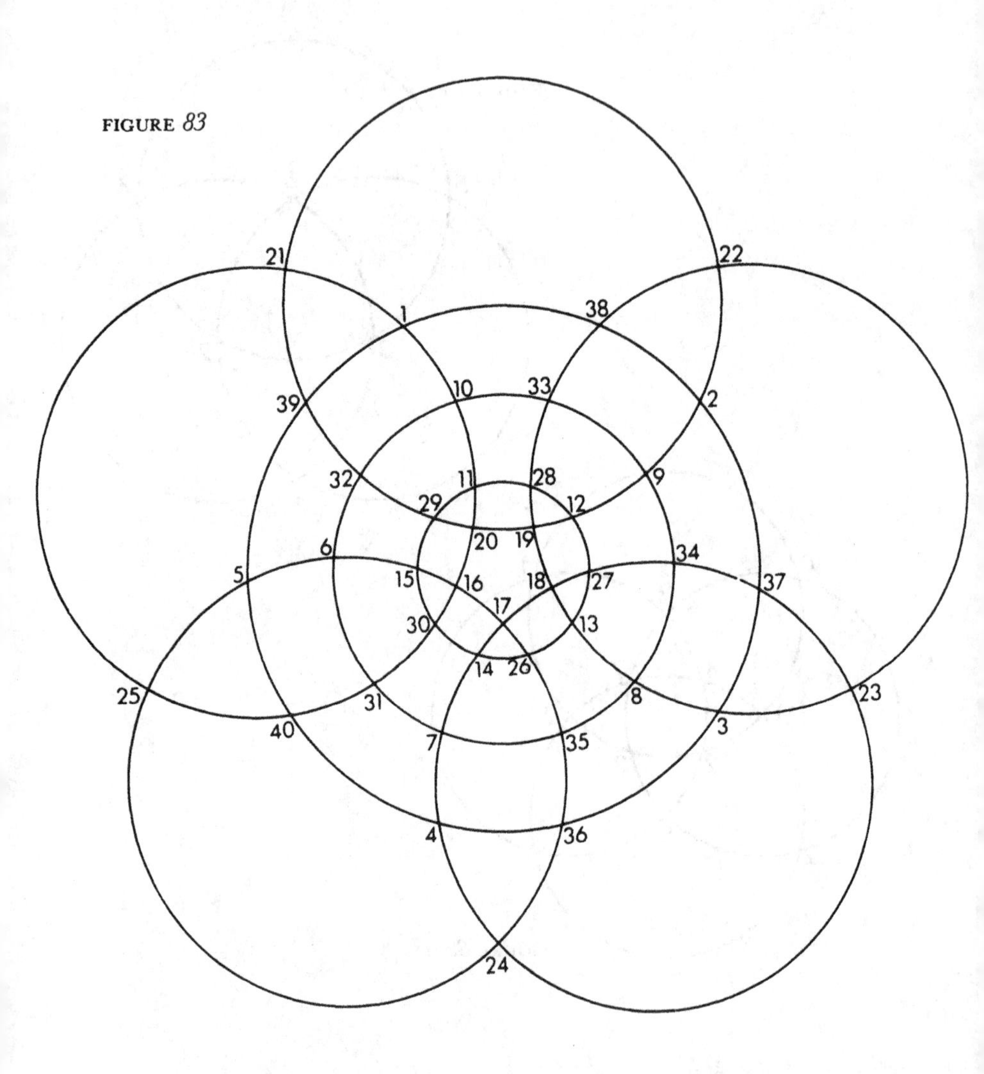

A simple five-pointed star can be used as the pattern for a magic arrangement of numbers; one is shown in Figure 84. The integers 1–12 are used with only 7 and 11 omitted. The magic constant for each straight line is 24 and is the smallest possible for this range of integers. There is no solution to this particular type of magic star if 10 consecutive integers are used.

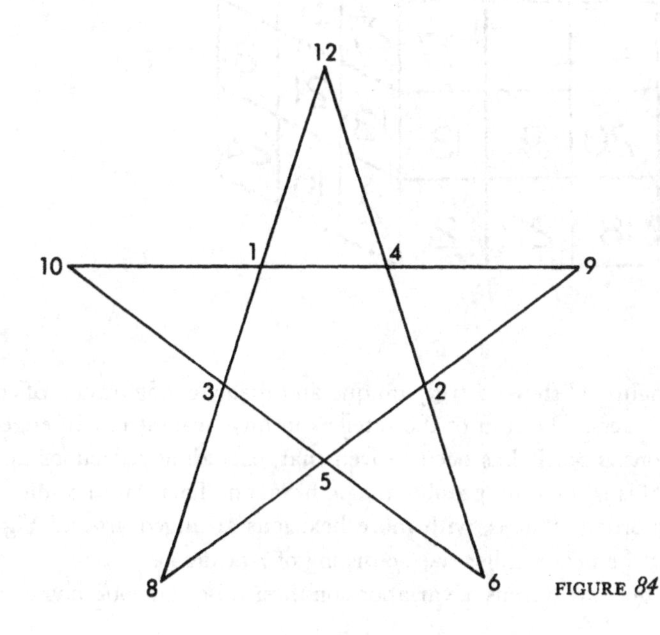

FIGURE *84*

Magic cubes are more difficult to devise, but Figure 85 is a simple example; it has a magic constant of 42. An $n \times n \times n$ magic cube consists of a series of integers so arranged that any column or row of n integers and the 4 main diagonals, each containing n integers, through the center of the cube add up to the same magic constant. Figure 86 shows the cube broken up into three component planes to show the internal structure. It should be noted that, with the exception of those in the central component plane, the main diagonals on the faces do not total 42.

FIGURE *85*

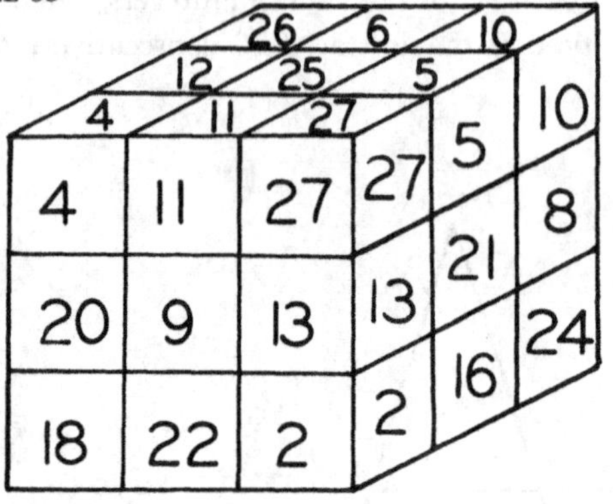

Figure 87 shows a truly unique and magic configuration of consecutive integers. The sum of the integers in any straight line of edge-joined hexagons is 38. It has been proven that, excluding reflections and rotations, this is the only possible magic hexagon. There are no solutions for higher orders, that is, with more hexagons arranged around Figure 87, nor for the next smaller size consisting of 7 hexagons.

There are dozens of variations on these miscellaneous magic arrange-

FIGURE *86*

4	11	27
20	9	13
18	22	2

12	25	5
7	14	21
23	3	16

26	6	10
15	19	8
1	17	24

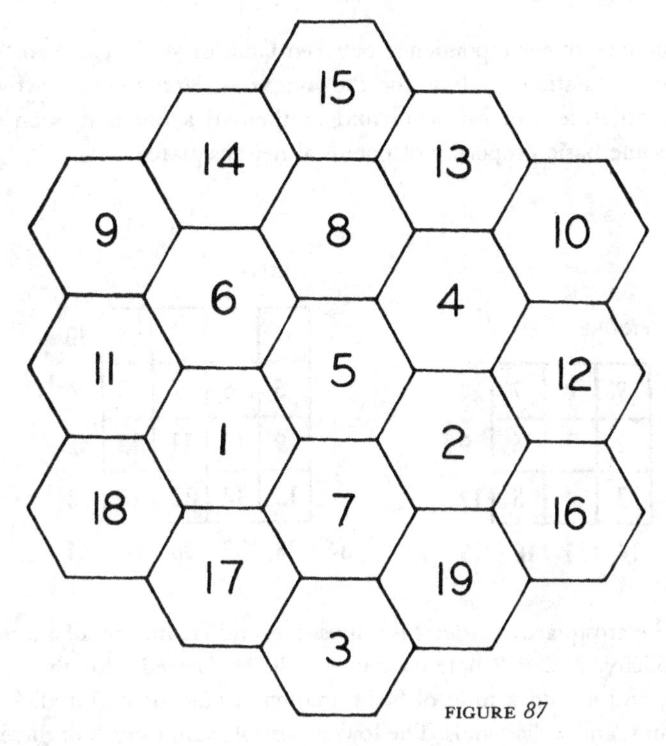

FIGURE 87

ments, and the possible configurations are limited only by one's ingenuity and imagination. Indeed, imagination is not a true limit, either—magic configurations in four dimensions have been constructed.

ANTIMAGIC SQUARES

In 1951, *Mathematics Magazine* asked for a proof that heterosquares of order 2 were impossible and also requested a heterosquare of order 3. A *heterosquare* was defined as an $n \times n$ array of the integers from 1 to n^2 such that all the rows, columns, and diagonals have *different* sums.

An exchange of correspondence between Charles W. Trigg, then "Problems and Questions" editor for the magazine, and the late Royal V. Heath, an American magician and mathematical puzzlist, soon established some basic properties of potential heterosquares.

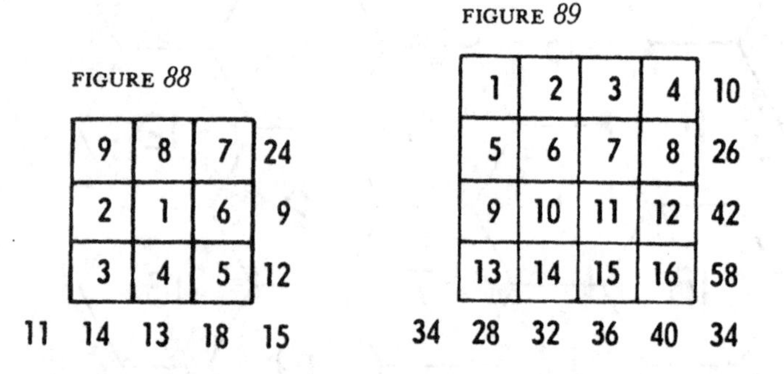

FIGURE *88*

FIGURE *89*

A heterosquare of order 2 is impossible, and Heath's proof is a marvel of simplicity. A 2×2 heterosquare would be formed with the integers 1, 2, 3, and 4, and a total of 6 summations would be required: 2 rows, 2 columns, and 2 diagonals. The lowest possible sum using 2 of these integers is $3 = 1 + 2$; the largest possible sum using 2 of these integers is $7 = 3 + 4$. This permits but 5 different sums, 3, 4, 5, 6, and 7, whereas 6 different sums are required. Heath thought that order 3 heterosquares were also impossible. However, Trigg found one, Figure 88.

There are at least two simple methods of producing heterosquares. Figure 89 shows the integers from 1 to 16 written across the rows of a 4×4 board. All the rows and columns have different sums, but the 2 main diagonals have the same sum, 34. A simple switch of the integers 15 and 16 produces the heterosquare shown in Figure 90.

Another method, already suggested by Figure 88, is a spiral technique. One starts from the center and places consecutive integers in a spiral, as shown in Figure 91.

Heterosquares, then, appear to be relatively easy to produce, and there is no particular beauty to the random values of the different sums. However, if these different sums must be *consecutive* integers, the problem becomes considerably more challenging. An *antimagic square* is defined as

FIGURE 90

	1	2	3	4	10
	5	6	7	8	26
	9	10	11	12	42
	13	14	16	15	58
34	28	32	37	39	33

FIGURE 91

	25	24	23	22	21	115
	10	9	8	7	20	54
	11	2	1	6	19	39
	12	3	4	5	18	42
	13	14	15	16	17	75
45	71	52	51	56	95	57

an $n \times n$ array of the integers from 1 to n^2 such that each row, column, and main diagonal produces different sums, and these sums are part of a consecutive series of integers. The following discussion of this amusement is almost entirely based on the work of J. A. Lindon, an English puzzlist and word-recreation enthusiast. Magic squares can be constructed by a variety of techniques, but there are virtually no systematic approaches to the construction of antimagic squares. The field is ripe for devoted and dedicated recreational mathematicians.

Analysis shows that for each order there are two, and only two, feasible series or sequences of numbers to be used for the sums. A simple way of obtaining these is as follows: for any given order, n, write down $2n + 1$ integers where the middle term is $\frac{n(n^2 + 1)}{2}$. This, as has been stated earlier, is the magic constant for an ordinary magic square of order n. These $2n + 1$ integers will be called the *generating sequence.* Then, the two feasible sequences for an antimagic square are those obtained by extending the generating sequence in either direction by one term. The

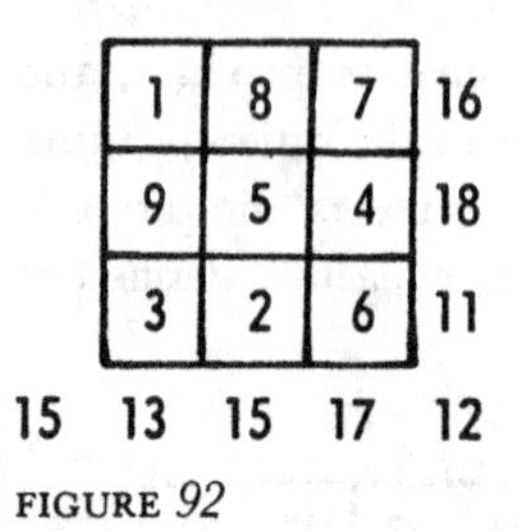

FIGURE *92*

sums of the diagonals will be equal to this additional term plus the middle term of the generating sequence.

The square shown in Figure 92 gives the totals 11, 12, 13, 15, 15, 16, 17, and 18, and is almost, but not quite, antimagic. Unfortunately, there is no 14, and 15 is repeated.

Antimagic squares of orders 1, 2, and, it is believed, 3 are impossible, but higher orders occur. Suppose a fourth-order antimagic square is to be constructed. The middle term of the generating sequence will be $(\frac{1}{2})(4)(4^2 + 1) = 34$, and the sequence will be 30, 31, 32, 33, 34, 35, 36, 37, and 38. So, in a fourth-order antimagic square, the 10 line totals must consist of one of the following: 29 to 38 inclusive, diagonals totaling $29 + 34 = 63$; or, 30 to 39 inclusive, diagonals totaling $39 + 34 = 73$.

Now, choose a sequence, abstract 2 terms having the correct total for diagonals, and separate the remaining $2n$ terms into two equal groups (having the same total) for rows and columns. There will be numerous permutations so mathematical analysis is difficult. Fortunately, trial and error can be used. The general method used by Lindon to find the examples of antimagic squares in this section of this chapter will be illustrated by completing the construction of a fourth-order antimagic square.

To double the chances of finding a result, choose 34 as one of the 2 diagonals, since this number can be associated with either 29 or 39. The

FIGURE *93*

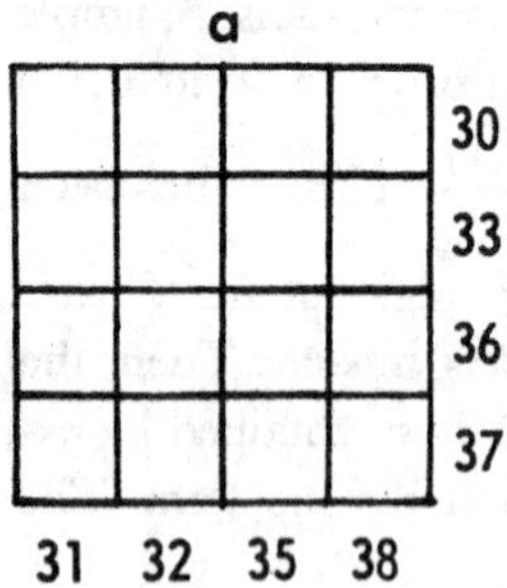

b

14	5	1	10
2	12	4	15
8	3	16	9
11	13	7	6

c

10	5	14	1
12	4	2	15
3	16	8	9
6	7	11	13

remaining 8 integers of the generating sequence can be separated into two groups: 30, 33, 36, 37 and 31, 32, 35, 38. Several other combinations are possible. Prepare a 4 × 4 blank diagram on a card, and number the rows and columns at the side as shown in Figure 93*a*.

Arrange, by trial and error, the numbers 1–16 so that the rows have the required totals. This is accomplished quickly and easily, and the result is shown in Figure 93*b*. Next, reshuffle horizontally until the columns are also right. This usually presents no trouble and the result is shown in 93*c*. Now test the diagonals. The 2 main diagonals have totals of 35 and 25, neither of which is of use. But any set of 4 numbers, one from each row and column taken from the square, can be rearranged as a principal diagonal, and several hopeful values are found among the 24 possibilities: in all, four 34's and a 39. Take each in turn and test it. In this way, the 39 gives, on rearrangement, the square shown in Figure 94, in which the other main diagonal has a total of 37. Unfortunately, 37 has already been used.

14	1	10	5	30
2	15	12	4	33
8	9	3	16	36
11	13	6	7	37
35	38	31	32	39

FIGURE *94*

	15	2	12	4	33
	1	14	10	5	30
	8	9	3	16	36
	11	13	6	7	37
34	35	38	31	32	39

FIGURE *95*

The situation is still not hopeless. It is usually possible to make various alterations in a square that leave a given diagonal unchanged. Give the bottom-left-to-top-right diagonal of Figure 94 its desired value by interchanging the two uppermost rows in their entirety. Substituting 2–1 for 14–15 drastically alters the 39-diagonal. Since 2 + 14 = 15 + 1, the horizontal interchanges 2 ↔ 15, 14 ↔ 1 can be made and so correct the

39 diagonal without altering the values of any row or column or that of the other diagonal. This antimagic square is shown in Figure 95.

The table below shows the 20 feasible, but not necessarily all the possible, combinations of numbers forming a fourth-order antimagic square sequence. All are possible except *B* and *L*, which fail for reasons of parity since the sums in all 4 rows or all 4 columns cannot be all even or all odd. The 4 rows or columns may of course be rearranged in any order and rows or columns interchanged. An example for each possible code letter for fourth-order antimagic squares is shown in Figure 96.

Sequence	*Diagonals*	*Rows*				*Columns*				*Code*
29 to 38	29/34	30	31	37	38	32	33	35	36	A
		30	32	36	38	31	33	35	37	B
		30	33	35	38	31	32	36	37	C
		30	33	36	37	31	32	35	38	D
	30/33	29	32	37	38	31	34	35	36	E
		29	34	35	38	31	32	36	37	F
		29	34	36	37	31	32	35	38	G
	31/32	29	33	36	38	30	34	35	37	H
		29	34	35	38	30	33	36	37	I
		29	34	36	37	30	33	35	38	J
30 to 39	34/39	30	31	37	38	32	33	35	36	K
		30	32	36	38	31	33	35	37	L
		30	33	35	38	31	32	36	37	M
		30	33	36	37	31	32	35	38	N
	35/38	30	31	36	39	32	33	34	37	O
		30	33	34	39	31	32	36	37	P
		30	33	36	37	31	32	34	39	Q
	36/37	30	32	35	39	31	33	34	38	R
		30	33	34	39	31	32	35	38	S
		30	33	35	38	31	32	34	39	T

FIGURE *96*

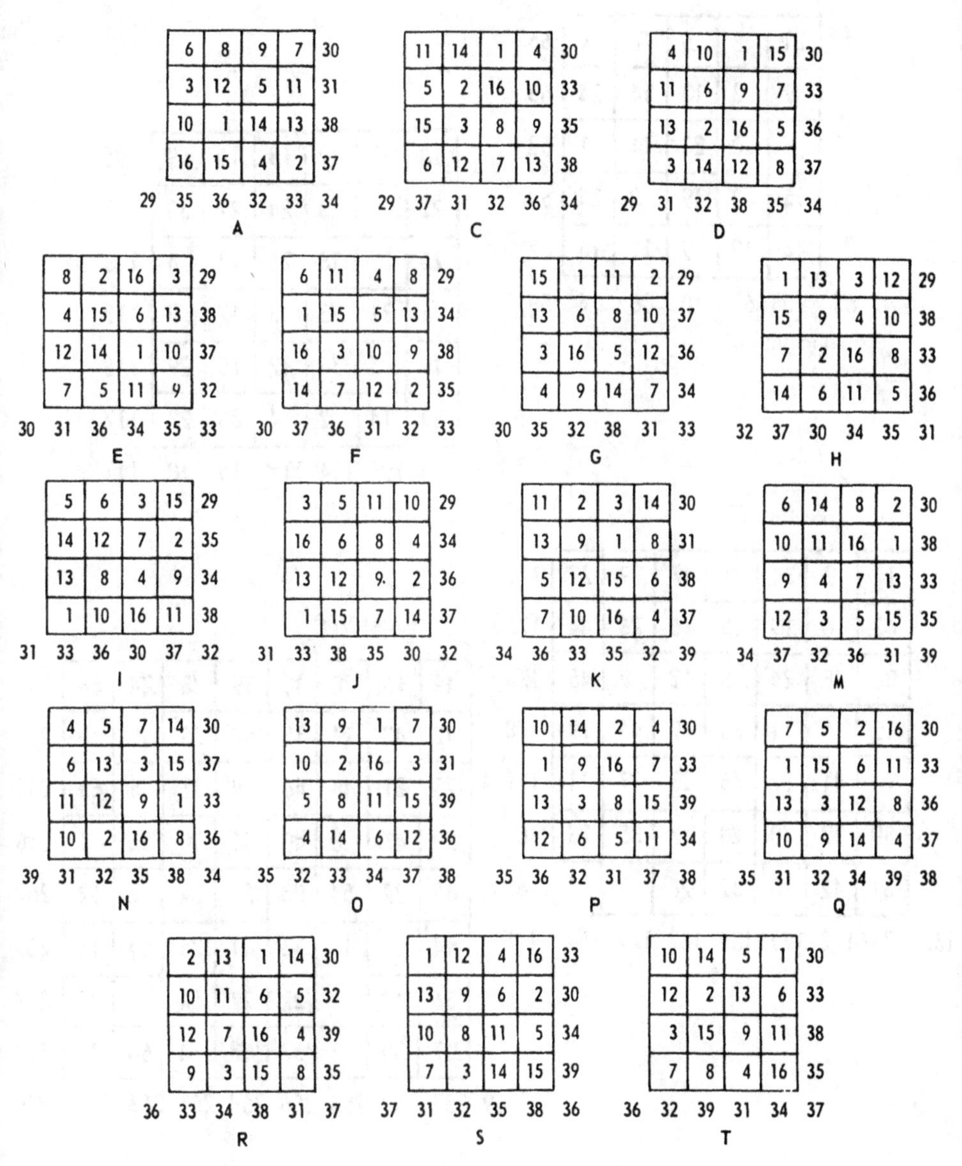

	21	18	6	17	4	66
	7	3	13	16	24	63
	5	20	23	11	1	60
	15	8	19	2	25	69
	14	12	9	22	10	67
65	62	61	70	68	64	59

FIGURE *97*

	10	25	32	13	16	9	105
	22	7	3	24	21	30	107
	20	27	18	26	11	6	108
	1	31	23	33	17	8	113
	19	5	36	12	15	29	116
	34	14	2	4	35	28	117
118	106	109	114	112	115	110	111

FIGURE *98*

FIGURE *99*

	14	3	34	21	47	29	22	170
	43	16	13	25	6	26	44	173
	30	48	24	8	12	9	45	176
	10	5	11	38	49	46	19	178
	4	41	37	36	33	27	1	179
	39	17	40	20	7	35	23	181
	31	42	18	32	28	2	15	168
183	171	172	177	180	182	174	169	175

FIGURE *100*

	49	16	50	10	19	28	24	56	252
	42	43	11	15	44	38	55	5	253
	25	21	48	46	9	37	6	63	255
	29	47	8	40	51	30	52	1	258
	45	22	54	23	20	34	2	62	262
	14	59	18	33	41	26	61	13	265
	36	12	58	32	27	64	3	35	267
	17	39	7	57	53	4	60	31	268
269	257	259	254	256	264	261	263	266	260

	52	19	81	22	29	15	42	31	76	367
	61	10	67	23	54	79	25	33	16	368
	57	9	71	24	38	1	51	47	75	373
	26	78	7	69	66	77	13	27	12	375
	39	21	74	20	37	17	49	55	64	376
	8	65	4	62	50	34	73	41	40	377
	56	68	2	63	14	72	35	44	6	360
	53	30	60	32	36	3	46	43	58	361
	11	70	5	59	48	80	28	45	18	364
379	363	370	371	374	372	378	362	366	365	369

FIGURE *101*

Figures 97–101 show examples of higher-order antimagic squares. As n increases, each square takes progressively longer to complete, but the first attempt will usually yield a result. With fourth-order antimagic squares, on the other hand, many failures occur.

This discussion must be regarded as no more than a preliminary skirmish with antimagic squares. Many areas of research are in need of study, including:

1. What are the general properties of antimagic squares?
2. Are there any systematic methods by which antimagic squares may be constructed? Such methods may bear some relation to the methods now available for the construction of magic squares. Perhaps the methods used to form heterosquares can be adapted to the construction of antimagic squares.
3. Can small antimagic squares be converted into larger ones?

4. Can magic squares be converted into antimagic squares by some routine manipulation of the integers? It would be interesting to find antimagic squares with properties comparable to the Franklin magic square shown in Figure 71.

5. Can antimagic squares be used in some practical manner? Magic squares have found application in the fields of marketing, agricultural, statistical, and atomic research.

TALISMAN SQUARES

Magic and antimagic squares have properties based on the sums of their rows, columns, and diagonals. Sidney Kravitz, a mathematician from Dover, New Jersey, ignored such considerations and searched for and studied other properties of number arrays. He devised the *talisman square,* which is defined as an $n \times n$ array of the integers from 1 to n^2 so that the difference between any one integer and its neighbor is greater than some given constant. A *neighboring integer* is defined as one immediately adjacent to a given number either vertically, horizontally, or diagonally. Figure 102 shows a talisman square in which the difference between any integer and its neighbor is greater than one. The integer 7, for example, has only 3 neighbors (3, 13, 15); 14 has 5 neighbors (12, 6, 4, 8, 16); and 13 has 8 neighbors (5, 3, 7, 15, 8, 4, 6, 11). Figure 103 shows a talisman square in which the differences are greater than 2.

FIGURE *102*

1	5	3	7
9	11	13	15
2	6	4	8
10	12	14	16

FIGURE *103*

5	15	9	12
10	1	6	3
13	16	11	14
2	8	4	7

The study of these squares is so new, in fact, that no rules for construction are known, nor are there any mathematical theories that can help in deciding the maximum possible difference between integers and their neighbors. In a 3 × 3 square containing the integers from 1 to 9, it is impossible to have a difference greater than 1 between all neighboring integers. The integer in the center is surrounded by the other 8 and, necessarily, one or two of these must be numerically consecutive to it.

FIGURE *104*

15	1	12	4	9
20	7	22	18	24
16	2	13	5	10
21	8	23	19	25
17	3	14	6	11

FIGURE *105*

28	10	31	13	34	16
19	1	22	4	25	7
29	11	32	14	35	17
20	2	23	5	26	8
30	12	33	15	36	18
21	3	24	6	27	9

Several other talisman squares of higher order were found by Kravitz. Figure 104 shows a 5 × 5 with differences greater than 4. In Figure 105, the 6 × 6 talisman square has differences greater than 8.

Talisman squares are only one aspect of this type of problem. Talisman hexagons, such as that shown in Figure 106, where the differences are greater than 4, can be devised also; compare this one with the magic hexagon of Figure 87. The problem can become more complicated if there are more than 8 possible neighbors. Figure 107 shows a grid of triangles in which one might try to insert the integers from 1 to 36 so that differences greater than 1 exist between any integer and its 12 neighbors. Triangle *A* touches 12 other triangles via 3 sides, and 3 other triangles at each vertex. Such construction might prove impossible unless the grid of triangles is larger.

As with antimagic squares there is room for much more study. Possible questions include:

1. What are the general properties of talisman squares?
2. Are there any systematic methods by which talisman squares, or rectangles, hexagons, and so on, can be constructed?

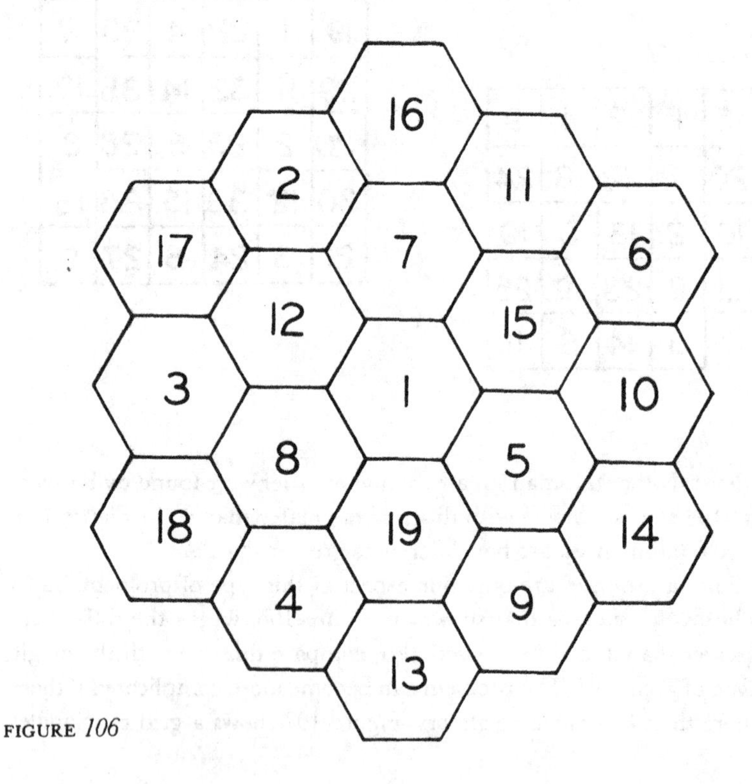

FIGURE *106*

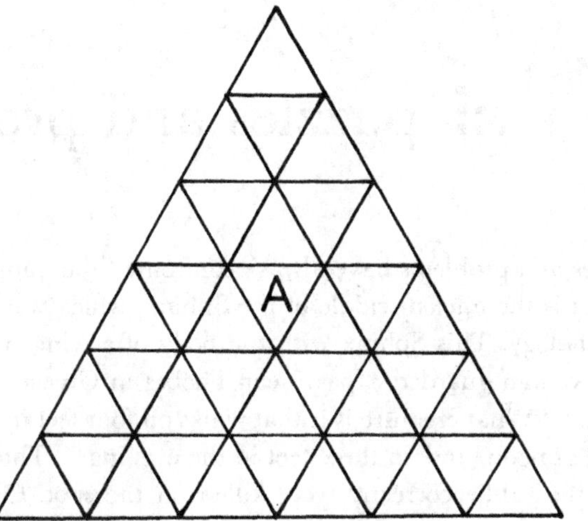

FIGURE *107*

3. Are there any talisman squares that are also magic or antimagic?

4. What is the maximum possible difference between integers and their neighbors for a given *n*th-order talisman square?

5. Can talisman squares find practical application? The agricultural scientist, for example, might find some use for them when separation of mutually interfering species of plants is desired on a restricted area of highly variable soil.

5

.:: puzzles and problems

Puzzles and problems have always confronted the human race. One of the oldest is the ancient riddle of the Sphinx, which was recounted in Greek mythology. This Sphinx with the body of a winged lion and the head of a woman guarded a pass near Thebes in Greece and asked all who went by: "What creature is it that walks on four feet in the morning, on two feet at noon, and on three feet in the evening?". Those who failed to answer the riddle correctly were killed on the spot. One passer-by, Oedipus, finally answered correctly by saying "Man crawls on all fours as a baby, then walks upright in his youth, and uses a staff in his old age." The Sphinx leaped from her pedestal and died. Oedipus later became the King of Thebes.

From that time to the present, humans have wrestled with enigmatic statements with amusement and earnestness. The 26 puzzles and problems below cover a rather full range. Puzzles in logic, numbers, measuring, geometry, probability, and geography have been mixed together. The answers will be found on pages 125–143.

PROBLEMS

1. MOONSHINE SHARING

Three distillers of illegal whisky wanted to share the 6 gallons of corn liquor they had made. The only measuring containers available were 3 cylindrical measuring cans holding 10, 11, and 13 quarts, respectively, and the barrel holding the 6 gallons of liquor. Feuding was imminent unless they could figure out how to share the liquor equally.

The cans were marked as to their full content and gave no indications of individual quarts. How did the moonshiners measure out 3 containers with 2 gallons in each and avoid feuding?

2. NUMBER TOUGHIES

What two whole numbers, neither containing any zeros, when multiplied together equal exactly 1,000,000,000?

If you solved that one, then find two whole numbers, again containing no zeros, that produce 1,000,000,000,000,000,000 when multiplied together.

3. NINE-COIN MOVE

Three dimes, 3 pennies, and 3 nickels are arranged in a single row in alternating order: nickel-penny-dime-nickel-penny-dime-nickel-penny-dime. Only 2 adjacent unlike coins can be moved at any one time to some other part of the line of coins. The moved coins must again be in contact with the line. Also, no coin pairs can be reversed: a penny-dime combination, for example, must remain in that order in its new position when it is moved. No pair of coins is to be moved and placed away from the remaining ones, but all coins need not touch one another. What is the minimum number of moves needed to get all the dimes, pennies, and nickels together to form: dime-dime-dime-penny-penny-penny-nickel-nickel-nickel?

4. SCOTCH AND WATER

The patient read the prescription that would save his life: "Mix carefully a 1-pint drink, made of scotch whisky and water, mixed 1 to 5 (1/6 scotch, 5/6 water). Drink it quickly and go to bed."

However, he found only the following items on hand:

A 1-quart bottle (32-ounce capacity) about half-full of scotch.

An 8-ounce glass.

An unlimited supply of water from his faucet.

A sink with a drain.

He had no other container or measuring device. He could pour from either container to the other without spilling a drop and could fill either to the brim without loss. How should he mix the prescription?

5. GEOMETRIC CONSTRUCTIONS

Construct the smallest equilateral triangle inscribable in a square.

Construct the largest equilateral triangle inscribable in a square.

Show, by simple means, that the triangles so constructed are indeed the smallest and largest possible. If the square is a *unit square,* that is, the sides are equal to 1, calculate the lengths of the sides of the inscribed triangles.

6. THE MAD HATTED!

Three persons are shown 3 red hats and 2 black hats. They are then seated in chairs placed in single file and blindfolded. A hat is placed on each person's head, the remaining hats hidden, and the blindfolds removed. One at a time, they are asked if each can guess the color of the hat on his own head.

The person who sits in the third chair is asked first, and he confesses that he does not know the color of his hat, even though he can see the hats on the heads of his 2 companions. The second person, who can see only the hat on the person in front of him, also admits he cannot guess his color. The first person, who can see no hats at all, says he knows the color of his hat and he is correct.

What color hat is he wearing, and how did he know?

7. HIGH STAKES

Mike sat down and started shuffling the cards. "What stakes?" he asked.

"Let's make it a gamble," Steve replied, putting a few bills and some coins on the table. "The first game, the loser pays 1 cent, the second, 2 cents, and so on. Double up each time."

"Okay," laughed Mike, checking his cash. "I've got only $6.01 and I'm not playing more than 10 games anyway."

So they played, and game followed game until at last Mike stood up. "That's my last cent I've just paid you," he declared, "but I'll have my revenge next week."

How many games had they played, and which did Mike win?

8. DEATH IN THE DECANTER

Messrs. Beer, Brandy, Chianti, Cider, Port, Sherry, and Whisky of the Bottle Club held their annual "booze-up," the drinks being beer, brandy, chianti, cider, port, sherry, and whisky. In conformity with the rules of the club, each samples exactly 3 different drinks, but nothing beginning with his own initial. After the party, 3 genuine corpses are found among the 7 drunks under the table, and investigation shows that one of the drinks had been maliciously tampered with. The following facts also come to light: (*a*) the 3 victims among them sampled all 7 drinks; (*b*) the brandy drinker was a married man who refused sherry; his name drink was drunk only by Mr. Brandy; (*c*) Mr. Whisky did not drink port; (*d*) of the 7, only Messrs. Port, Sherry, and Whisky have never married; (*e*) the port drinkers outnumbered the whisky drinkers by one; (*f*) the namesake of the only drink sampled by all 3 bachelors drank chianti; Mr. Chianti himself was the only other to try the drink all 3 bachelors had sampled; (*g*) the murderer did not touch the name drink of any of his victims.

Who poisoned whom, and which was the fatal tipple?

9. PROBLEMS IN PROBABILITIES

What is the probability that a number chosen at random from the range 1,000,000,000 to 9,999,999,999 will contain 10 different digits?

What is the probability that 10 different digits will occur in a 10-digit sequence written at random?

10. THE GOLDEN SPHERES

"Goldsmith, with this great pile of glittering gold coins you are to make for me this night 4 golden spheres," ordered the Pharaoh.

"I have consulted the priests in this matter, and the radius of each sphere, measured in groods, must be a whole number smaller than 12; and each sphere must be a different size. The sum of the diameters of the smallest and largest spheres must be 26 groods. This rule applies also to the sum of the other 2 diameters. Furthermore, you must use every scrap of gold. Now set to work. If you do not finish the work exactly as I have specified by the rising of the sun tomorrow, you will die."

As soon as the Pharaoh left, the trembling craftsman began feverishly to work by his furnace, weighing and melting gold and making calculations. He finally arrived at a solution correct in all respects but one: the largest sphere had a radius of 12 groods. In desperation, the goldsmith decided to risk the Pharaoh's anger. He completed the work just before the light of day. At dawn he was executed for failure to follow the exact instructions of the priests.

What sizes should the spheres have been and what sizes were the spheres the goldsmith had constructed?

11. A PAPER-COVERING PROBLEM

The object is simple: cover as much of a piece of paper 8½″ × 12½″ as possible with seven 3″ × 5″ file cards. Overlapping of the cards and even overlapping the edges of the paper is allowed, but the cards cannot be cut or folded in any way.

12. THE COMMONER'S DILEMMA

The King did not like the idea of the young commoner and his romantic attachment to his only daughter. However, the young man refused to stay away from her, and in order to get rid of him the King summoned the young man.

"You scoundrel, you won't stay away from my daughter. Now, I'm a fair man, and though I could have you beheaded immediately for disobeying my orders, I'm going to give you a chance to marry my daughter and live happily ever after. Are you willing to gamble your life?" The young man, very much in love, accepted, and the King explained. "Tomorrow, I will assemble the whole court; and in their presence you will draw out of a bag, held by me, one of two peas which I will place therein. One of the peas will be white, and, if you draw it, you can marry my daughter with my blessing. The other pea will be black and if you draw *that* one you will be beheaded on the spot."

The next day the King assembled his court and very slyly slipped two black peas into the bag. The Princess happened to see her father and just managed to signal her lover of the nefarious plot. Unfortunately, the King spoke before the young man could protest.

"Well, young man, I see you showed up! As you can see I brought the Royal Headsman for an immediate execution, if necessary. If you decline to draw, or try to talk your way out of this, I'll have you beheaded for insulting my daughter by implying she isn't worth the risk."

The poor fellow was certainly in a quandary. He could not refuse to draw, nor could he accuse the King of being a villain. He had to choose a pea from the bag, and the pea was bound to be black.

As it turned out, the King commended the young man for his cleverness; the marriage took place with all the pomp possible under the circumstances, and the newlyweds received a small kingdom of their own to rule.

What happened?

13. DID THE BUTLER DO IT?

"Where are those valuable Indian-head pennies I left on the table this morning, James? I put them in a square array and now there are only 2 left. You didn't take them, did you?"

"Well, sir," replied the butler, "shortly after you left, three burglars came in. They divided the pennies equally among themselves, but left these 2 because they could not divide them equally."

Is James telling the truth?

14. AN AIRPORT PROBLEM

A young man about to take off in his plane spotted a pretty young girl on the concourse.

"Hi, did you miss your plane?" he asked.

"Yes, and now I'm stranded here for a few more days."

"I'd be glad to give you a lift," he offered.

"But," she replied, "you don't even know where I'm going!"

"It doesn't matter. I can take you there without going out of my way more than a few miles."

Naturally, the girl thought this was a fresh young man with a new line and refused his offer until he told her where he was headed. She realized he had been telling her the truth and went with him.

Now, where was the young man going?

15. THE SEVEN FORTUNES

Seven children each owned a different number of pennies. The ratio of any one's fortune to the fortune of every child poorer than himself was an integer. In total, the combined fortune of the children was 2,879 pennies.

How many pennies did each have?

16. OCCUPATIONAL MIX-UP

Eustace has four friends whose occupations are butcher, baker, tailor, carpenter and whose names are Mr. Butcher, Mr. Baker, Mr. Tailor, and Mr. Carpenter, though the names do not correspond to the occupations. Each man has a son and a daughter, but no son practices the same trade as his father, nor does any practice a trade corresponding to his own name.

Each son marries a daughter whose maiden name does not suggest her husband's or his father's trade. Each girl changes her last initial when she marries.

The baker's son married Miss Butcher. Mr. Butcher, Sr. is not a baker. The trade of Mr. Carpenter, Sr. is the same as young Mrs. Butcher's maiden name.

Eustace managed to figure out the trade of each of the 8 males and the maiden names of the sons' wives. Can you?

17. A TRAVELING MAN

A traveler sets out to cross a desert. On the first day, he covers $\frac{1}{10}$ of the journey; on the second day, he goes $\frac{2}{3}$ of the distance traveled the first day. He continues on in this manner, alternating the days on which he travels $\frac{1}{10}$ of the distance still to be covered, with days on which he travels $\frac{2}{3}$ of the total distance already covered. At the end of the seventh day he finds that $22\frac{1}{2}$ miles more will see the end of his journey.

How wide is the desert?

18. A PROBLEM IN CONFUSION

When Bert was just one year younger than Bill was when Ben was half as old as Bill will be 3 years from now, Ben was twice as old as Bill was when Ben was $\frac{1}{3}$ as old as Bert was 3 years ago. But, when Bill was twice as old as Bert, Ben was $\frac{1}{4}$ as old as Bill was one year ago.

Ignoring odd months and considering that Bert has passed the half-century mark, it will be no problem to find out how old these three friends are.

19. CIGARETTE SELLING

A cigarette factory sells cigarettes in packs of 20 and of 43 cigarettes. 63 cigarettes could be sold without breaking open any packs, but not 64 or 65. 189 cigarettes can be sold without breaking open any packs. In the same way, 558, 559, 560, or 561 cigarettes can be sold in whole packs, but packs would have to be broken to sell 557 cigarettes.

What is the largest number, N, of cigarettes that cannot be sold without breaking open any packs? Any number of cigarettes greater than N can be sold using whole packs only.

What is the general solution to this problem for any 2 packs of different size consisting of L and M cigarettes?

20. EIGHT STAMPS

Three prisoners are to be given a chance at freedom. The Warden brings a paper bag to their cell and tells them that in the bag are 3 red stamps and 5 green stamps. Blindfolded, the prisoners are to reach into the bag, pick out 2 stamps, and paste them on their own foreheads. The blindfolds will then be removed and each man will be allowed to look at the other prisoners but is not to communicate in any way with them.

If the first man, A, cannot deduce the color of the stamps on his own forehead, then the second man, B, will be asked. If he cannot deduce the color of the stamps on his forehead, C will be asked. If C does not know the color of his stamps, then A will be asked again. This will continue, in order, until one of the men can deduce the color of the stamps on his forehead, and that prisoner will be set free.

Now it so happens that each prisoner pulls out a red and a green stamp, so that when the blindfolds are removed, each man sees 2 red and 2 green stamps.

Which prisoner, A, B, or C, first knows *absolutely* the color of his own stamps, and in which round of questioning?

21. SQUARED EGGS

Thomas has an egg tray the size of a small table top with square partitions like a blank crossword puzzle diagram and completely filled with brown and white eggs that are all the same size. Collecting together the brown eggs, Tom is delighted to find that he has just enough to fill a perfect square in the center of the tray, leaving a uniform margin of white eggs all around. On the other hand, when he assembles the white eggs in the exact center of the tray, he finds they also make a perfect square, similarly oriented, one size larger than before, but only provided he puts a single brown egg in the very middle.

How many eggs of each color has he?

22. THE ORACLE OF THE THREE GODS

A certain oracle is presided over by three gods, who take turns in answering the questions put to it by pilgrims. The three gods are the God of Truth, who always tells the truth; the God of Falsehood, who always lies; and the God of Equivocation, who alternately tells the truth and lies.

One day a pilgrim arrives and wishes to know whether or not his wife is faithful. Unfortunately, the poor pilgrim does not know which god will answer his question. Moreover, the priests of the oracle allow no more than one question from each pilgrim.

How should the pilgrim state his question, so as to be sure of either his wife's fidelity or unfaithfulness?

23. CUBE FORMATION

What is the shortest strip of paper 1-inch wide and black on one side that can be folded to form a 1-inch cube that is black on all sides?

24. BALANCING CUBES

The twins, Mary and Tony, had received a kitchen scale with 2 large balancing pans and a set of 12 wooden blocks, all different in size and color. The largest block was 12 inches on a side, the next 11 inches, and so on by inches down to the smallest block, which was only 1 inch on a side. The blocks were colored silver, gold, crimson, yellow, violet, white, black, emerald, indigo, orange, purple, and red, though not necessarily in order according to size.

The twins went outside to play, and, after tiring of building towers and pyramids, they began balancing the blocks on the scale. It was surprising how many balances they were able to achieve. After a while they started recording the balances as they found them. Just before a sudden downpour forced them back into the house they had noted the following balances:

1. The yellow and black blocks balanced the gold and emerald blocks.
2. The purple block alone balanced the red, silver, and violet blocks.
3. The black, orange, purple, and violet blocks balanced the emerald, gold, indigo, silver, white, and yellow blocks.
4. The gold block alone balanced the indigo, purple, and yellow blocks.

After the storm, the twins discovered the blocks had been completely washed clean, and they begged to have them repainted as they were colored originally.

Match the original color and corresponding size of each of the 12 wooden blocks.

25. MAJOR PERKINS

Assuming the truth of the following statements, what can you deduce about Major Perkins?

Only bleary-eyed military men ever try to shave with a toothbrush.

No bleary-eyed military men are at all musical.

Men who never try to shave with a toothbrush never play golf.

Major Perkins plays the tuba on Wednesdays.

Only musical people who are golfers forget to change their socks.

26. THE CHICKEN YOLKS

A certain odd breed of chickens, the "Queerchick," lays eggs that may contain 1, 2, or no yolks. Farmer Jones used to keep track of the number of queerchick eggs he and his family ate. He stopped counting when he reached exactly 5,000 eggs and also lost the records he had kept of the number of yolks in each. However, he did recall that most of the 5,000 eggs had only 1 yolk and exactly ½ of the remaining eggs had 2 yolks.

How many yolks were there in those 5,000 eggs?

ANSWERS

1. MOONSHINE SHARING

This solution hinges on the fact that the 10-quart cylindrical container can be half emptied by tipping it until the overflow at the top of the can is level with the first appearance of the bottom of the can. This is the same as cutting an upright cylinder in half by passing a plane through its upper right edge and lower left edge. Thus, the steps are first to fill the 13-quart container; pour the liquor from it into the 10-quart can, leaving 3 quarts in the first container. Half empty the 10-quart container into the barrel, so no whisky will be wasted, and pour the 5 quarts remaining in the 10-quart can into the 11-quart one. Add the 3 quarts remaining in the 13-quart can into the 11-quart container. There are now 8 quarts in the 11-quart container for the first moonshiner.

Repeat all the above steps, but pour the 3 quarts from the 13-quart can into the 10-quart can, which contains 5 quarts. There are now 8 quarts in the 10-quart can for the second moonshiner. The barrel holds the remaining 8 quarts, which can be poured into the 13-quart can for the third moonshiner.

Lloyd Jim Steiger, of Vallejo, California, found a solution that did not involve tipping any containers, utilizing instead a technique that is uncommon for this type of problem. Fill the 13-quart container leaving 11 quarts in the barrel. Fill the 10-quart can from the 13-quart one and pour the 3 quarts remaining in the latter into the 11-quart container. Now pour the 10 quarts from the 10-quart can into the 13-quart one, and fill the rest of the cân from the whisky still remaining in the barrel. This leaves 8 quarts in the barrel. Pour the 3 quarts that are now in the 11-quart can into the 10-quart one and place the 10-quart container into the 11-quart container. Fill the 11-quart can from the 13-quart one; with the 10-quart can (with 3 quarts of liquor in it and, thereby, taking up 3 quarts of volume) floating in the 11-quart can, only 8 quarts can be put into the 11-quart container. The 3 quarts that are in the 10-quart container can now be added to the 5 quarts still remaining in the 13-quart can, making a total of 8 quarts in it.

It must be pointed out, however, that such "floatation" methods work only if the containers being floated, the 10-quart one in this case, have no weight. Otherwise, a quantity of liquor equal in weight to the weight of the container would be displaced and, in the case described, less than 8 quarts could be poured into the 11-quart can.

2. NUMBER TOUGHIES

These problems can be solved in a surprisingly simple manner if it is remembered that 1,000,000,000, for example, is 10 multiplied by itself 9 times. In mathematical shorthand, this is written with the exponent 9,

1,000,000,000 = 10^9. Since 10 = (2)(5), then $10^9 = [(2)(5)]^9 = (2^9)(5^9)$. $2^9 = 512$, and 5^9 = 1,953,125. Therefore:

$$1{,}000{,}000{,}000 = (512)(1{,}953{,}125)$$

Similarly:

$$1{,}000{,}000{,}000{,}000{,}000{,}000 = 10^{18} = (2^{18})(5^{18})$$
$$= (262{,}144)(3{,}814{,}697{,}265{,}625)$$

The problem is a fascinating one, and several puzzlists have gone to great lengths to find other solutions: namely, what other powers of 10, besides 9 and 18, have 2 factors containing no zeros? The easiest way to find such solutions is to find all values of 2^n containing no 0's and then find the corresponding values of 5^n having no 0's. Examination of mathematical tables (the *Handbook of Mathematical Tables* published by the Chemical Rubber Publishing Co.) shows that the following powers of 2 less than 100 contain no 0's: 1, 2, 3, 4, 5, 6, 7, 8, 9, 13, 14, 15, 16, 18, 19, 24, 25, 27, 28, 31, 32, 33, 34, 35, 36, 37, 39, 49, 51, 67, 72, 76, 77, 81, and 86. Only a few of these are matched by corresponding powers of 5 containing no 0's:

$$10^1 = (2)(5)$$
$$10^2 = (2^2)(5^2) = (4)(25)$$
$$10^3 = (2^3)(5^3) = (8)(125)$$
$$10^4 = (2^4)(5^4) = (16)(625)$$
$$10^5 = (2^5)(5^5) = (32)(3{,}125)$$
$$10^6 = (2^6)(5^6) = (64)(15{,}625)$$
$$10^7 = (2^7)(5^7) = (128)(78{,}125)$$
$$10^9 = (2^9)(5^9) = (512)(1{,}953{,}125)$$
$$10^{18} = (2^{18})(5^{18}) = (262{,}144)(3{,}814{,}697{,}265{,}625)$$
$$10^{33} = (2^{33})(5^{33}) = (8{,}589{,}934{,}592)(116{,}415{,}321{,}826{,}934{,}814{,}453{,}125)$$

Rudolph Ondrejka of Atlantic City, New Jersey checked all the values of 2^n up to $2^{5{,}000}$ and found no powers of 2 greater than 86 that did not contain zeros. Fred Gruenberger, a mathematician at the RAND Corporation in Santa Monica, California, programmed an IBM 1620 electronic computer to produce consecutive powers of 2 and type the exponent for those that lacked zeros. The machine dutifully typed the values listed above. It went to $2^{10,535}$ before the computer had to be cut off. $2^{10,535}$ is a number with 3,172 digits. These digits pass the frequency test of randomness, so there are about 317 each of 0's and 5's in the number. If a number containing n digits passes the frequency test of randomness it means that each of the 10 digits, 0, 1, 2, 3, 4, 5, 6, 7, 8, 9, appears approximately 10 per cent of the time in the given number. Roughly half of the 0's and 5's will produce a new 0 on the next power, and the empirical results indicate that the probability of a power of 2 higher than 86 having no 0's is quite low. The matter awaits an analytical proof, but the above analysis tends to discourage further empirical searching.

For those who are interested in further empirical results, T. Charles Jones, a student at Davidson College in Davidson, North Carolina, reports that up to $2^{57,134}$ (which has 17,200 digits) the results are equally negative. And, I can add, even *if* a power of 2 greater than 86 *is* found which contains no 0's, the probability that the corresponding power of 5 would also contain no 0's is fantastically small.

3. EIGHT-COIN MOVE

Eight. Let N = nickel, P = penny, and D = dime. The moves are:

Move	*Order of Coins*												
0			N	P	D	N	P	D	N	P	D		
1	N	P	N	P	D	N	P	D			D		
2	N			P	D	N	P	D			D	P	N
3	N	P	D			N	P	D			D	P	N
4	N	P	D	D	P	N	P	D					N
5	N	P	D	D			P	D	P	N			N
6	N	P	D	D	D	P	P			N			N
7			D	D	D	P	P			N	N	P	N
8			D	D	D	P	P	P	N	N	N		

It is interesting that no nickel-dime moves are made. Nor does the transposition of coins change the position on the line of coins as a whole.

4. SCOTCH AND WATER

The prescription can be filled in 10 steps. (1) Fill the 8-ounce glass with scotch. (2) This is the most painful step! Pour the rest of the scotch down the drain. (3) Pour the scotch from the glass into the bottle. (4) Fill the glass with water. (5) Empty the glass into the bottle (the bottle now holds 16 ounces of a 50–50 mixture). (6) Fill the glass from the bottle. (7) Throw the contents of the glass down the drain (another painful step). The bottle now holds 8 ounces of a 50–50 mixture. (8 and 9) Put 2 glasses full of water into the bottle (it now holds 24 ounces of a mixture consisting of 20 ounces of water and 4 ounces of scotch). (10) Fill the glass from the bottle. The 16 ounces remaining in the bottle is the required prescription.

5. GEOMETRIC CONSTRUCTIONS

All 3 vertexes of an equilateral triangle inscribed in a square must touch the sides of the square or coincide with its corners. This limits the possibilities to two:

Case 1. The 3 vertexes of the triangle are on 3 sides of the square.
Case 2. One vertex of the triangle is in a corner of the square and the other vertexes are on the 2 sides of the square opposite this corner.

For Case 1, 2 vertexes of the triangle must be on opposite sides of the square; that is, on either both vertical sides or both horizontal sides. Minimum-length sides are achieved only if the side of the triangle formed by these 2 vertexes is parallel to a side of the square, as shown in Figure 108*a*, where line *EF* is parallel to *AB*. If the side of the square is unity, then the sides of the triangle are also unity. Construction of this triangle is simple: bisect *CD* at *G*. With *G* as center and radius *AB* strike arcs

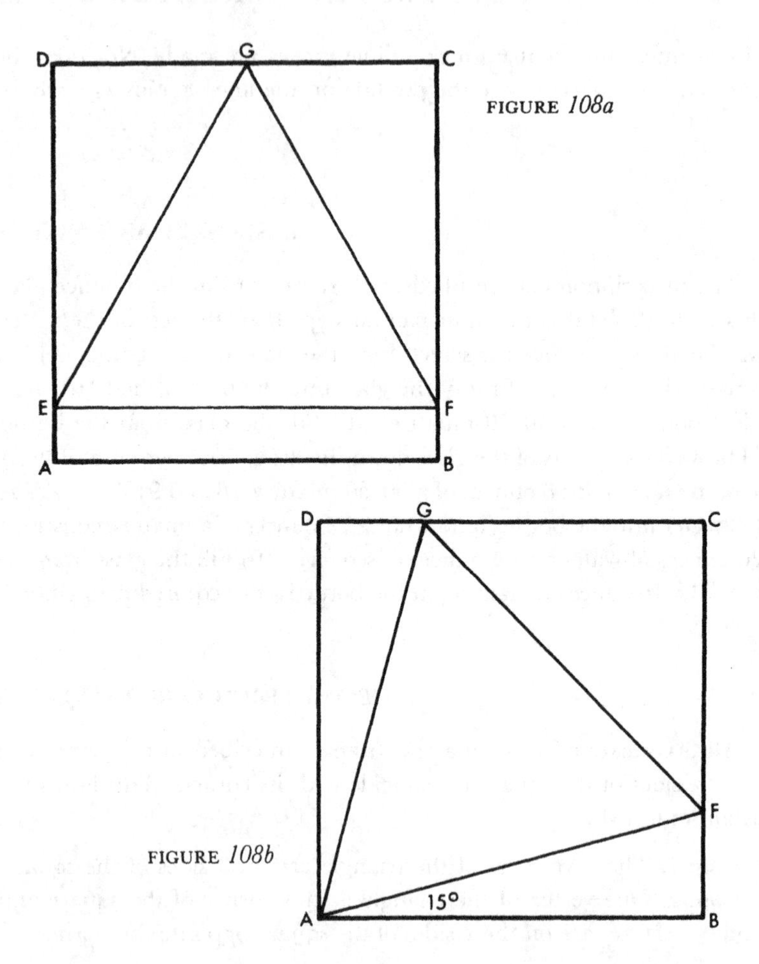

FIGURE *108a*

FIGURE *108b*

through AD and BC at E and F. Triangle EFG is the required equilateral triangle. $EA = FB = \frac{1}{2}(2 - \sqrt{3})$.

For Case 2, with a vertex of the triangle in a corner of the square, AF may range in length from 1 (AB) to approximately 1.0353. The triangle is constructed by first making a 15° angle at A. (An equilateral triangle, constructed for Case 1, has angles equal to 60°. If one of these angles is bisected, two 30° angles are produced. Bisection of one of these 30° angles produces the required 15° angle.) Use length AF for the lengths of the other 2 sides of the triangle. If the angle FAB is greater than 15°, an equilateral triangle could not be constructed in the square such that the other 2 vertexes of the triangle touch the 2 sides of the square opposite the vertex at A. The Case 2 solution is shown in Figure 108*b*.

So, the smallest equilateral triangle inscribable in a unit square has sides equal to unity. The largest such equilateral triangle in a unit square has sides equal to approximately 1.0353.

6. THE MAD HATTED!

Let the men be called Jack, John, and Joe. Joe sees Jack and John. John can see only Jack. If Joe saw 2 black hats he would know his own hat was red since there were only 2 black hats available. Since Joe says he does not know the color of his own hat, then he must see either 2 red hats or a red hat and a black hat. John logically concludes from Joe's negative response exactly what Joe sees. If John saw Jack with a black hat then he (John) would know his hat was red. Since he says he does not know the color of his own hat, then he must see a red hat on Jack. Jack, who has been reasoning in the same manner, correctly deduces that the color of his own hat must be red.

7. HIGH STAKES

601 is less than 2^{10} but more than 2^9, so they played 10 games. The gross total of stakes was $2^0 + 2^1 + 2^2 + 2^3 + 2^4 + 2^5 + 2^6 + 2^7 + 2^8 + 2^9 = 1{,}023$ cents. If the gross winnings were: Steve x cents and Mike y cents, then $x + y = 1{,}023$ and $x - y = 601$ cents; solving, $x = 812$ and $y = 211$. Only the stakes in the first, second, fifth, seventh, and eighth games (1 cent, 2 cents, 16 cents, 64 cents, and 128 cents) add up to 211 cents. So Mike won the first, second, fifth, seventh, and eighth games.

8. DEATH IN THE DECANTER

Mr. Whisky poisoned Messrs. Brandy, Cider, and Port by putting something in the whisky. However, the wording of the puzzle does not rule out the possibility of the murderer having committed suicide and being among the victims. There would then be several possible solutions; among them, Mr. Beer committed suicide by poisoning the whisky, killing Messrs. Brandy and Cider.

9. PROBLEM IN PROBABILITIES

There are, for example, 6 ways of arranging 3 different digits; 24 ways of arranging 4 different digits; 120 ways of arranging 5 different digits. A short way of calculating the number of ways in which n digits can be arranged is to write $n!$, which means the product of all the integers from 1 to n. $4! = (1)(2)(3)(4) = 24$.

There are 10!, or 3,628,800, different ways of arranging the 10 distinct digits. The first problem gave the range from 1,000,000,000 to 9,999,999,999. Of the 3,628,800 ways of arranging the 10 digits, exactly 10 per cent start with a 0 and these cannot be included as acceptable arrangements. Only $\frac{9}{10}(10!)$ or $(9)(9!)$ arrangements are acceptable in the range chosen. The total number of integers in the range is $8{,}999{,}999{,}999 + 1 = (9)(10^9)$. So the probability required is

$$\frac{(9)(9!)}{(9)(10^9)} = \frac{(9!)}{(10^9)} = \frac{362{,}880}{1{,}000{,}000{,}000} = 0.00036288$$

That is, about 1 arrangement out of every 2,750 contains the 10 distinct digits.

The second problem is concerned with the full range from 0,000,000,000 to 9,999,999,999 (that is, *any* 10-digit sequence was specified, not only those from 1,000,000,000 on). This range comprises 10^{10} different numbers; 10!, or 3,628,800 of them will be those which include the 10 distinct digits. The probability, then, is $\frac{10!}{10^{10}}$ or $\frac{9!}{10^9}$, which is the same probability as in the first problem.

10. THE GOLDEN SPHERES

The volume of a sphere is $\frac{4\pi r^3}{3}$, where r is the radius. Let the total volume of the 4 spheres which the goldsmith was to have constructed be V. Then

$$\frac{4\pi A^3}{3} + \frac{4\pi B^3}{3} + \frac{4\pi C^3}{3} + \frac{4\pi D^3}{3}$$

or

$$\frac{4\pi}{3}(A^3 + B^3 + C^3 + D^3) = V$$

Since $\frac{4\pi}{3}$ and V are numerical constants, it is only necessary to consider the radii of the spheres. Tabulating those integer pairs that total 13: (1, 12); (2, 11); (3, 10); (4, 9); (5, 8); (6, 7). Since the goldsmith's fatal attempt used all the gold, the combination of radii utilized the listed pairs such that the sum of the 4 cubes of one set equals the sum of the 4 cubes of the other set. Only the following solution is possible:

$$2^3 + 11^3 + 3^3 + 10^3 = 1^3 + 12^3 + 5^3 + 8^3 = 2{,}366$$

The goldsmith's attempt must have involved spheres with radii of 1, 5, 8, and 12 groods. The correct solution should have been spheres with radii of 2, 3, 10, and 11 groods.

11. A PAPER-COVERING PROBLEM

This guileless little puzzle turned out to be full of the devil. After its first appearance in *Recreational Mathematics Magazine* in 1961, the solution of $99\frac{15}{16}$ square inches by the originator, Jack Halliburton of Los Angeles, California, was improved slightly by a number of puzzlists who managed to cover exactly 100 square inches. The placement is shown in Figure 109.

Martin Gardner posed the problem in 1962 in the column "Mathematical Games," which he edits in *Scientific American*, and a slightly better solution was found by Stephen Barr of Woodstock, New York, the author

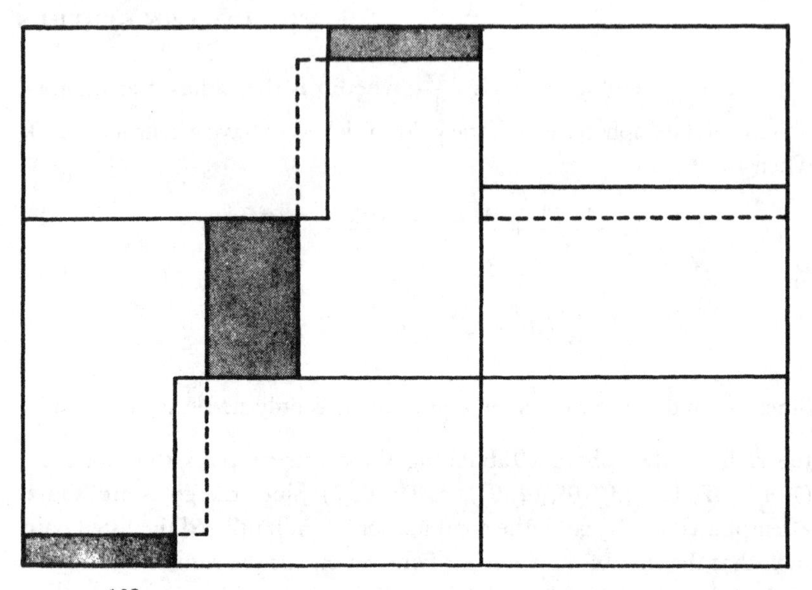

FIGURE *109*

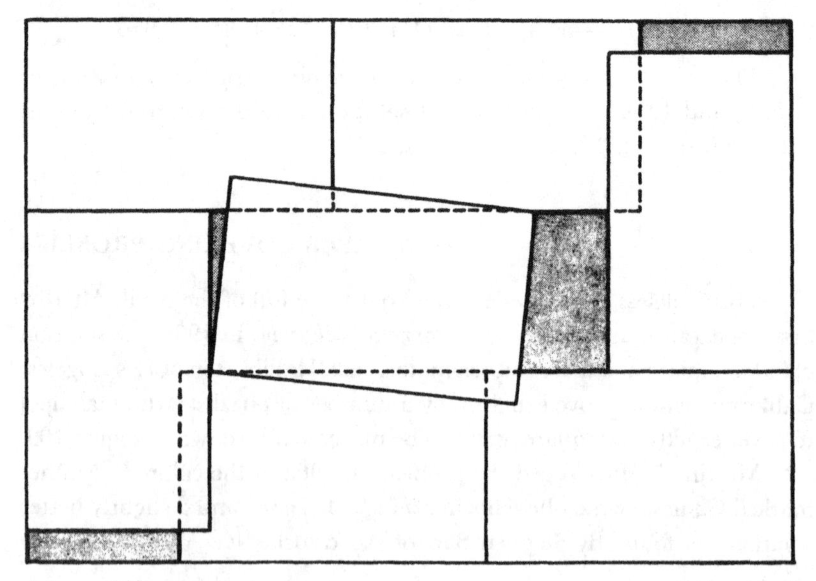

FIGURE *110*

of several books of mathematical recreations. By tilting one of the cards (Figure 110) at an angle of 5°34′32″, Barr managed to cover an area of 100.059+ square inches. However, it turned out that this was not the optimum solution, which was calculated as an angle of 6°12′30″, yielding a covering of 100.06844+ square inches. Gardner published this last solution in 1963.

The problem did not die there, however. R. Robinson Rowe of Sacramento, California, performed 3 different sets of calculations, correcting the 1963 value and arriving at another optimum angle. Rowe calculated a covering of 100.065834427+ square inches for the 6°12′30″ angle of tilt and found a better one: 6°12′37.8973″, which produces a covering of 100.065834498+ square inches. This is an improvement of 0.000000071 square inches.

Is there a better solution?

12. THE COMMONER'S DILEMMA

The solution to this apparently insoluble dilemma is amazingly simple. Announcing that he was taking white, our clever commoner reached into the bag, closed his fist around one pea, and quickly put it into his mouth and swallowed it. Naturally, everyone was curious and wondered which one he had chosen. The young man asked the King to show the pea that remained in the bag. It was black, of course, and the courtiers concluded that he had picked out the white pea. The King could not very well divulge his own nefarious plot, and so everyone lived happily ever after.

13. DID THE BUTLER DO IT?

The butler lied when he said that three burglars left 2 pennies after dividing the rest evenly. There is never a remainder of 2 when a square number is divided by 3. All whole numbers may be placed in one of three categories: $3k$, $3k + 1$, or $3k + 2$, where k is any whole number. Their squares are $9k^2$, $9k^2 + 6k + 1$, and $9k^2 + 12k + 4$, respectively. The first leaves no remainder when divided by 3, while the other 2 leave a remainder of 1.

14. AN AIRPORT PROBLEM

Since the young man could not even know the direction in which the girl was heading, one logical conclusion might be reached: he was not going anywhere in particular, and he was indeed handing the girl a line. This would be quite unacceptable, so the conclusion is reached that he did have some particular destination in mind. The only destinations that could be reached by going in any direction and still not put him "out of his way" would be the opposite side of the earth, *or* the very airport at which the conversation took place, that is, the young man may have been starting a round-the-world flight.

If a round-the-world flight was being planned, then any one of hundreds of airports in an English-speaking country could be a possible location. However, if a trip to the opposite side of the earth was involved, the ideal solution would be that which yielded 2 such English-speaking airports located on opposite sides of the earth. The following pair of locations fills the requirements: Perth, Australia located at 31°47′ south latitude and 116°0′ east longitude, and the Bermuda Islands located at 32°20′ north latitude and 64°55′ west longitude.

15. THE SEVEN FORTUNES

Let the first fortune be A; the second fortune, AB; the third, ABC; the fourth, $ABCD$; and so on. Then:

$$A + AB + ABC + ABCD + ABCDE + ABCDEF + ABCDEFG = 2{,}879$$

or

$$A(1 + B + BC + BCD + BCDE + BCDEF + BCDEFG) = 2{,}879$$

2,879 is a prime number and has only the 2 factors 1 and 2,879, so $A = 1$.

Then

$$B(1 + C + CD + CDE + CDEF + CDEFG) = 2{,}878 = (2)(1{,}439)$$

1,439 is also a prime, so $B = 2$. Similarly, $C = 2$, $D = 2$, $E = 2$, and $F = 2$. Finally, $F(1 + G) = 178 = (2)(89)$, so that $G = 88$.

The fortunes of the 7 children are 1, 2, 4, 8, 16, 32, and 2,816 pennies, respectively, making a total of 2,879 pennies.

16. OCCUPATIONAL MIX-UP

Mr. Carpenter, Sr., is a tailor, and his son is a butcher who married Miss Baker. Mr. Baker, Sr., is a butcher, and his son is a tailor who married Miss Carpenter. Mr. Tailor, Sr., is a baker, and his son is a carpenter who married Miss Butcher. Mr. Butcher, Sr., is a carpenter, and his son is a baker who married Miss Tailor.

17. A TRAVELING MAN

Calculation yields $\frac{3x}{16}$ miles remaining on the seventh day, where x is the width of the desert. Since $\frac{3x}{16} = 22.5$ miles, $x = 120$ miles.

18. A PROBLEM IN CONFUSION

Bert is 72; Ben is 80; Bill is 85.

19. CIGARETTE SELLING

The greatest number of cigarettes that cannot be delivered without opening packs containing 20 and 43 cigarettes is 797. The general solution for 2 packs of different sizes consisting of L and M cigarettes is $N = LM - (L + M)$, when L and M have no common factor other than unity. For example, if $L = 5$ and $M = 10$ both have a common factor, 5; whereas, if $L = 5$ and $M = 12$, there is no common factor except unity, and $L = 5, M = 12$ would be a valid solution, yielding $N = 60 - 17 = 43$.

20. EIGHT STAMPS

Prisoner *B*, in the second round of questioning, is the first to be able to state correctly the color of the stamps on his forehead. His reasoning is as follows:

B thinks to himself, when his turn comes up in the second round: I see 2 reds and 2 greens. Therefore, I can have either red-green or green-green on my own forehead. *If* I have green-green, *A* would have thought to himself on his second turn:

> I see 3 greens and one red, so I can have red-red, red-green, or green-green. But if I had red-red, *B* would have known in the first round that he had green-green, since all the reds would have been visible to him. So I cannot have red-red. Suppose I had green-green. Then *C*, in the first round, would have seen 4 greens and known that he (*C*) could only have red-green or red-red. So *C* would have reasoned to himself:
>
> > If I had red-red, *B* would know that he (*B*) has green-green from *A*'s answer that he (*A*) did not know the color of his own stamps. But *B* did not know he has green-green, so I cannot have red-red. Therefore, I must have red-green.
>
> However [says *A*], *C* did not conclude he had red-green, so I (*A*) cannot have green-green. I have red-green.

However [thinks *B*], *A* did not know at the beginning of the second round that he had red-green, therefore I cannot have green-green. I must have red-green.

After laboriously explaining his reasoning, *B* was given his freedom. He deserved it.

21. SQUARED EGGS

Figure 111 shows the problem. Let b and w represent the number of brown and white eggs, respectively, and n^2 the total number of eggs in the tray. Let p be the number of eggs along each side of the brown inner square and q the corresponding number for the white square. Clearly, since these squares are central, the next size larger must have 2 eggs more per side. Then

$$w = q^2 - 1 = (p + 2)^2 - 1 = p^2 + 4p + 3$$
$$b = p^2$$
$$w + b = n^2$$

Then $$n^2 = 2p^2 + 4p + 3$$

Or $$n^2 - 2x^2 = 1, \qquad \text{where } x = p + 1.$$

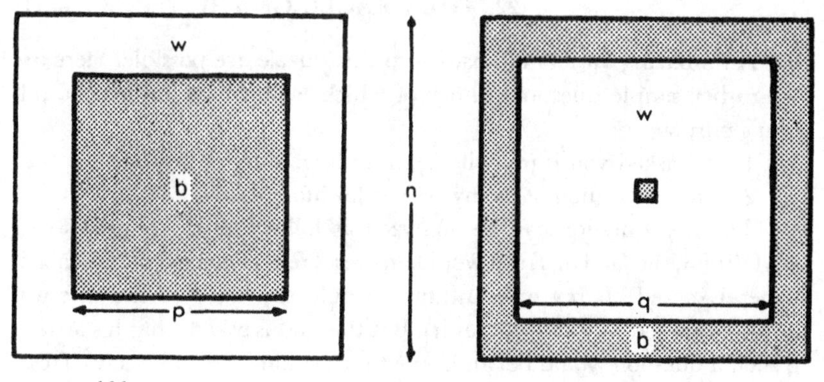

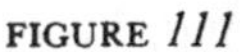
FIGURE *111*

This indeterminate equation has to be solved in integers and there are an infinite number of solutions. The first 3 are $(n, x) = (3, 2); (17, 12); (99, 70)$. The first does not apply to the problem, and the third would require considerably more than a "small table top." The second yields $p = 11$, giving the answer as 121 brown eggs and 168 white eggs.

There is a slight flaw in the statement of this problem: the species of eggs is not mentioned. On this basis, Diane S. Bitsch and Mike McDonald, both of St. Louis, Missouri, noted that the choice of a particular solution from the infinite set is subject to the further constraint that the eggs be of such a size as to be capable of simultaneous positioning upon a table top that is small. The following list was submitted to suggest additional solutions:

Brown	*White*	*Species of Eggs*
121	168	Chicken
4,761	5,040	Pigeon
165,649	167,280	Caviar

22. THE ORACLE OF THE THREE GODS

An amazing variety of answers to this puzzle are possible. Here are two rather simple questions either of which would have served the pilgrim's purpose.

1. If I asked you if my wife is faithful, would you say yes?
2. Are you as honest as my wife is faithful?

The first question can be analyzed as follows. If the pilgrim's wife was faithful, the God of Truth would answer "Yes." The God of Falsehood, if merely asked "Is my wife faithful?" would say "No" (if the wife was faithful, this god would lie about it). But this god is asked what his answer to such a question would be, and, true to his calling, he falsely says "Yes." The God of Equivocation, no matter whether it is his turn to tell the truth or to tell lies, would likewise answer "Yes" to the pilgrim's question.

In other words, if the answer to the first question is yes, then the pilgrim's wife is faithful. Conversely, if the pilgrim's wife is unfaithful, all the gods would answer no to the question.

A similar analysis for the second question yields the same result: A yes by the gods means the wife is faithful, a no means she is not.

23. CUBE FORMATION

No more than 8 inches of 1-inch-wide paper are required. Figure 112 shows the solution.

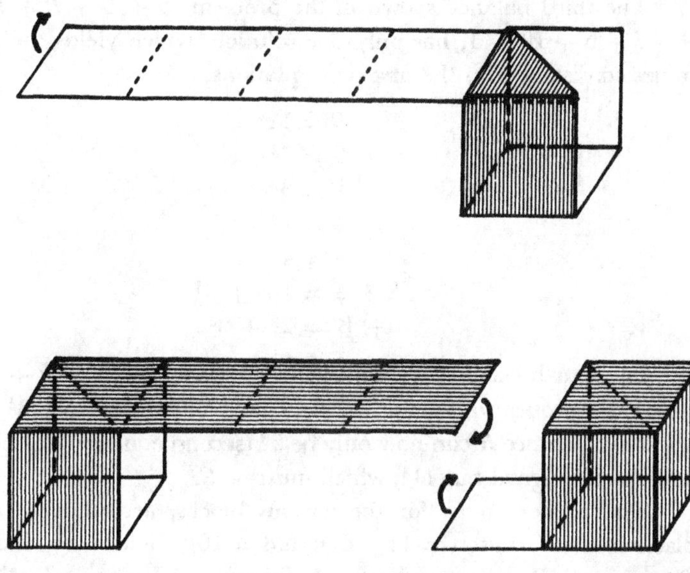

FIGURE *112*

24. BALANCING CUBES

Let black be B, crimson C, emerald E, gold G, indigo I, orange O, purple P, red R, silver S, violet V, white W, and yellow Y. The first equation, $Y + B = G + E$, has only one solution involving the sums of cubes:

$$1^3 + 12^3 = 9^3 + 10^3$$

There are 3 possible solutions to the second equation, $P = R + S + V$:

$$12^3 = 6^3 + 8^3 + 10^3$$
$$9^3 = 1^3 + 6^3 + 8^3$$
$$6^3 = 3^3 + 4^3 + 5^3$$

The third of these equations is the only one consistent with the first cube balance given. So $P = 6^3$; that is, the 6-inch cube was purple.

The third balance shown in the problem, $B + O + P + V = E + G + I + S + W + Y$, has only one solution, which yields the following values consistent with the first two equations:

$$B = 12^3$$
$$O = 7^3$$
$$V = 3^3$$
$$Y = 1^3$$
$$S = 4^3$$
$$E + G = 10^3 + 9^3$$
$$I + W = 2^3 + 8^3$$

The fourth balance shown in the problem, $G = I + P + Y$, determines the proper values for E, G, I, and W as 10^3, 9^3, 8^3, and 2^3, respectively. Since R can now only be 5^3 (second equation), all the values have been assigned but 11^3, which must be C.

The proper colors for the various blocks, according to size, are: Black = 12; Crimson = 11; Emerald = 10; Gold = 9; Indigo = 8; Orange = 7; Purple = 6; Red = 5; Silver = 4; Violet = 3; White = 2; Yellow = 1.

Robert S. Johnson of Town of Mt. Royal, Quebec, pointed out that all the cubes could have been used to form a single balance:

$$Y + W + S + I + G + B = V + R + P + O + E + C$$

that is,

$$1^3 + 2^3 + 4^3 + 8^3 + 9^3 + 12^3 = 3^3 + 5^3 + 6^3 + 7^3 + 10^3 + 11^3$$

25. MAJOR PERKINS

Many puzzlists who submitted solutions to this logic problem included an assumption that was not, strictly speaking, entirely valid. It was assumed that Major Perkins was a man. If so, the logical deduction from the statements in the problem is that he is musical, not bleary eyed, not a golfer, does not try to shave with a toothbrush, and he remembers to change his socks.

The statements of the problem do not entirely justify such a rash assumption. Major Perkins *could* have been a female Major, in the Salvation Army. Actually, the only completely valid conclusion that can be reached from the statements as given is that Major Perkins (sex not determined) plays the tuba on Wednesdays.

26. THE CHICKEN YOLKS

There were exactly 5,000 yolks in the 5,000 eggs.

6

.:: number recreations

Any book on mathematical recreations must eventually come to number amusements. I hope here to avoid repetition of often-published number oddities. Familiar material is examined from a somewhat different viewpoint, and a generous selection of new work or results is also included.

NUMBERS AND THEIR DIVISORS

Any number can be divided by any other number. The results of such division are either whole numbers or fractions that can be expressed as exact or repeating decimals. 94 can be divided by 47 with a result of 2; 94 can also be divided by 5, and the answer is 18⅘ or 18.8; or, when 94 is divided by 7, the result is 13.428571428571 . . . , with 428571 repeated indefinitely.

In practice, however, recreational mathematicians are usually interested only in those divisors of a number that do not result in fractions or decimals, but, rather, in whole numbers. In the examples above, only the division by 47 produced a whole number, 2. The number 94 can also be divided by 1 and by 94 to produce the whole numbers 94 and 1, respectively. Hence, in this sense, the divisors of 94 are 1, 2, 47, and 94. Still another restriction is usually made on the divisors; the number itself is excluded as a divisor. Then, 94 has only the divisors 1, 2, and 47. Therefore, the divisors of a number, N, will be defined as any whole numbers smaller than N that, when divided into N, produce whole numbers; such divisors are called *aliquot divisors*. The aliquot divisors of 28 are 1, 2, 4, 7, and 14; those of 284 are 1, 2, 4, 71, and 142; there is only 1 aliquot divisor for 17 and that is 1. Further reference to divisors will mean aliquot divisors unless otherwise indicated.

Interest in the aliquot divisors of a number centers on what happens when they are added together to produce another number, N_1. If the aliquot divisors of N_1 are added together, another number, N_2, will result. If the aliquot divisors of N_2 are added together, another number, N_3, will result. If this process is continued indefinitely, then 1 of 5 end results will occur.

1. If the sum of the divisors of N is equal to N, N is defined as a *perfect number*. The divisors of 28 are 1, 2, 4, 7, and 14, and their sum is 28, so 28 is a perfect number. If the sum of the divisors of N is equal to a whole-number multiple of N, N is defined as a *multiperfect* number (the divisors of 120 are 1, 2, 3, 4, 5, 6, 8, 10, 12, 15, 20, 24, 30, 40, and 60, and their sum is 240). There are other numbers whose divisor sums are equal to 3 or 4 or more times the numbers themselves. Tradition, however, has encumbered mathematicians with a slightly different definition of multiperfect numbers than given here for perfect numbers. The distinction will be brought out when these numbers are treated in more detail.

2. If the sum of the divisors of N is equal to N_1, and the sum of the divisors of N_1 is equal to N, N and N_1 are called *amicable numbers*. For example, the divisors of 220 are 1, 2, 4, 5, 10, 11, 20, 22, 44, 55, and 110, and their sum is 284. The divisors of 284 are 1, 2, 4, 71, and 142, and their sum is 220. Then 220 and 284 are amicable numbers; they are sometimes referred to as an amicable pair of numbers.

3. If, after more than 2 steps in the process of successive additions of the divisors of a number, the original one is obtained, the number is called a *sociable number*. The divisors of 12,496 are 1, 2, 4, 8, 11, 16, 22,

44, 71, 88, 142, 176, 284, 568, 781, 1,136, 1,562, 3,124, and 6,248; and their sum is 14,288. There are 19 divisors of 14,288, and their sum is 15,472; 15,472 has 9 divisors, and their sum is 14,536; 14,536 has 15 divisors, and their sum is 14,264; and 14,264 has 7 divisors, and their sum is 12,496. So, after 5 steps, the original number of 12,496 is reached. The sociable number 14,316 repeats itself after 28 steps. Over a dozen other sociable numbers are known, all of which repeat after 4 steps.

4. If the sum of the divisors of N is equal to N_1, and the process is continued indefinitely, obtaining the divisors of N_1, then N_2, then N_3, and so on to N_n, it may be that N_n has only one divisor, 1. Then N_n is a *prime* number, which is defined as an integer that has only the divisor 1. The divisors of 22 are 1, 2, and 11, and their sum is 14; the divisors of 14 are 1, 2, and 7, and their sum is 10; the divisors of 10 are 1, 2, and 5, and their sum is 8; the divisors of 8 are 1, 2, and 4, and their sum is 7; the divisor of 7 is 1. Hence, 7 is a prime number. Of course, the series of divisor-additions may start with a prime number, and the process will stop with the first step.

5. If the numbers resulting from each successive divisor-addition step continually increase, the end result, never obtained, is infinity. Unfortunately, it is not known whether any such series actually exists. If one starts with 138, one obtains 150 by summing the divisors of 138. The divisors of 150 total 222, those of 222 total 234, and so on. After 120 more steps the integer reached is 8,698,040,672. Finally, after 177 such steps, the series terminates with 1. However, for the number 276, the number 208,751,080,955,844 is reached after 75 steps. Further work was not possible with the computer used.

If much work is to be done in these studies of divisors and their sums, it is best to adopt a standard notation. The symbol $s(N)$ means the sum of the aliquot divisors of N. If $s(N) = N$, then N is a perfect number. If $s(N) = N_1$ and $s(N_1) = N$, then N and N_1 are amicable numbers. A modification of this symbol can be used to carry the series of necessary steps indefinitely. For instance, the sociable number 12,496 can be represented in its series of divisor-additions as $s^5(12{,}496) = 12{,}496$, which means that 12,496 is reached after 5 successive additions of the divisors of the numbers starting with 12,496. In the same manner, the sociable number 14,316 can be represented as $s^{28}(14{,}316) = 14{,}316$. Amicable numbers, then, can be represented as $s^2(N) = N$.

Although computers have been used for work on prime and perfect numbers, the amicable and sociable numbers, until recently, have been left to the human computer, who is limited in both speed and accuracy. In any case, the number 276, which yielded 208,751,080,955,844 after 75 steps, *may* produce a continually increasing series of divisor-additions, in which case it could be represented by $s^{\infty}(276) = \infty$. If 276 proves to be a sociable number, it could be represented by $s^k(276) = 276$, where k is the number of steps required. The only other possibility is that 276 may eventually yield a prime number, thus ending the series with 1. The number 22 produces the prime 7 after 4 steps; thus $s^4(22) = 1$.

To summarize, every continued series of aliquot divisor-additions, sometimes called aliquot suites, produces either a perfect number, a multiperfect number, an amicable pair, a prime number, a sociable number, or, possibly, infinity. Since only a few sociable numbers are known, and no examples of the production of infinity are known, it is only necessary here to deal with perfect, multiperfect, amicable, and prime numbers.

PERFECT AND MULTIPERFECT NUMBERS

It has already been stated that if $s(N) = N$, then N is a perfect number. Another way of saying this is that a perfect number is an integer, the sum of whose aliquot divisors is equal to the integer itself. The first perfect number is 6, since $1 + 2 + 3 = 6$. The second perfect number is 28, since $1 + 2 + 4 + 7 + 14 = 28$. The next 3 perfect numbers are 496; 8,128; and 33,550,336. Until 1952, only 12 perfect numbers were known, and the problem was at a level impractical for desk calculators. Then SWAC, the National Bureau of Standards' Western Automatic Computer, found the next 5 perfect numbers. More were found subsequently and, by 1971, 24 perfect numbers were known.

All even perfect numbers are of the form $2^{p-1}(2^p - 1)$, where p and $(2^p - 1)$ are primes. The primes of the form $2^p - 1$ are called *Mersenne primes* and are named after the French mathematician Marin Mersenne (1588–1648) who, in 1644, announced a list of new perfect numbers. Although his list was far from perfect (there were 5 errors) the achievement was magnificent for his time. His remarkable work is honored by calling all numbers of the form $2^p - 1$, where p is a prime, Mersenne numbers and symbolizing them as M_p. Those Mersenne numbers that are also primes are called Mersenne primes. For all Mersenne primes there is a corresponding even perfect number.

The terms 2^p and 2^{p-1} are read as "two to the p power" and "two to the $p - 1$ power." As was noted earlier, this means that 2 is multiplied together p or $p - 1$ times:

$$2^3 = (2)(2)(2) = 8 \qquad 2^7 = (2)(2)(2)(2)(2)(2)(2) = 128$$

The only known values of p that yield Mersenne primes and corresponding perfect numbers are: 2, 3, 5, 7, 13, 17, 19, 31, 61, 89, 107, 127, 521, 607, 1,279, 2,203, 2,281, 3,217, 4,253, 4,423, 9,689, 9,941, 11,213, and 19,937. Though there are undoubtedly many more beyond $p = 19{,}937$, the size of these primes grows rapidly as p increases: the eighth Mersenne prime ($p = 31$, giving the prime $2^{31} - 1 = 2{,}147{,}483{,}647$) has 10 digits. The fifteenth Mersenne prime, $(2^{1,279} - 1)$, has 386 digits; and the largest known Mersenne prime (the largest known prime, in fact) has 6,002 digits.

The perfect numbers are products of a Mersenne prime and a power of 2, and are correspondingly larger than the Mersenne primes. The twenty-fourth Mersenne prime has 6,002 digits, as noted above, but the twenty-fourth perfect number, which is $2^{19,937}(2^{19,937} - 1)$, has 12,003 digits. The table below gives the short form (using powers of 2) of the 24 known perfect numbers, the number of digits in each, and the number of aliquot divisors. All the perfect numbers known are *even*, and it is not even known whether any odd perfect numbers can exist. Only the full values of the first 7 perfect numbers are shown below; the others become so large that at least 12 additional book pages would be required to print them out. Perfect numbers are traditionally represented by the letter V.

Perfect Number	*Number of Digits*	*Number of Aliquot Divisors*
$V_1 = 2^1(2^2 - 1) = 6$	1	3
$V_2 = 2^2(2^3 - 1) = 28$	2	5
$V_3 = 2^4(2^5 - 1) = 496$	3	9
$V_4 = 2^6(2^7 - 1) = 8{,}128$	4	13
$V_5 = 2^{12}(2^{13} - 1) = 33{,}550{,}336$	8	25
$V_6 = 2^{16}(2^{17} - 1) = 8{,}589{,}869{,}056$	10	33
$V_7 = 2^{18}(2^{19} - 1) = 137{,}438{,}691{,}328$	12	37
$V_8 = 2^{30}(2^{31} - 1)$	19	61
$V_9 = 2^{60}(2^{61} - 1)$	37	121
$V_{10} = 2^{88}(2^{89} - 1)$	54	177
$V_{11} = 2^{106}(2^{107} - 1)$	65	213
$V_{12} = 2^{126}(2^{127} - 1)$	77	253
$V_{13} = 2^{520}(2^{521} - 1)$	314	1,041
$V_{14} = 2^{606}(2^{607} - 1)$	366	1,213
$V_{15} = 2^{1,278}(2^{1,279} - 1)$	770	2,557
$V_{16} = 2^{2,202}(2^{2,203} - 1)$	1,327	4,405
$V_{17} = 2^{2,280}(2^{2,281} - 1)$	1,373	4,561
$V_{18} = 2^{3,216}(2^{3,217} - 1)$	1,937	6,433
$V_{19} = 2^{4,252}(2^{4,253} - 1)$	2,561	8,505
$V_{20} = 2^{4,422}(2^{4,423} - 1)$	2,663	8,845
$V_{21} = 2^{9,688}(2^{9,689} - 1)$	5,834	19,377
$V_{22} = 2^{9,940}(2^{9,941} - 1)$	5,985	19,881
$V_{23} = 2^{11,212}(2^{11,213} - 1)$	6,751	22,425
$V_{24} = 2^{19,937}(2^{19,937} - 1)$	12,003	39,873

A multiperfect number is traditionally defined as an integer whose divisors, *including* the integer itself, add up to an integral multiple of the integer. That is, $s(N) + N = kN$, where k is some whole number greater than 1. The first multiperfect number (discovered by Mersenne, who did so much for perfect numbers) is 120; it has the aliquot divisors 1, 2, 3, 4, 5, 6, 8, 10, 12, 15, 20, 24, 30, 40, and 60, which total 240. But the definition of a multiperfect number demands the inclusion of the integer itself, 120 in this case, as one of the divisors; the sum of *all* the divisors of 120 is $240 + 120 = 360$, which is exactly 3 times 120. Therefore, 120 is called a *tri*perfect number. This can be represented as $s(120) + 120 = (3)(120) = 360$. All perfect numbers are really *bi*perfect numbers, but the traditional difference between the definitions of perfect and multiperfect numbers automatically excludes the inclusion of perfect numbers in the latter group.

The largest known multiperfect number has 264 digits, compared with the 12,003 digits of the largest known perfect number. There are 550 multiperfect numbers known, ranging from triperfects to octoperfects (the sum of all the divisors is equal to 8 times the integer itself).

Class	*Number of Multiperfects Known*
Triperfect	6
Quadriperfect	36
Quinqueperfect	62
Sexiperfect	222
Septiperfect	217
Octoperfect	7

There is no space to list all 550 multiperfect numbers here, but the 6 triperfect numbers are: 120, 672, 523,776, 459,818,240, 1,476,304,896, and 31,001,180,160.

As pointed out, all known perfect numbers conform to the formula, $2^{p-1}(2^p - 1)$, but there is no known rule or formula that will yield multiperfect numbers. In some respects, multiperfect numbers have more fascination than perfect ones. The huge size of the larger perfect numbers precludes any "hand work." On the other hand, Alan L. Brown, one of the last of the "human computers," of Montclair, New Jersey, has devoted more time and energy to the study of multiperfect numbers than anyone, and has tabulated all of them. A number of these were discovered by him *without* the aid of electronic computers.

Much work remains to be done with perfect and multiperfect numbers. Among the many questions still requiring answers are: Are there any more multiperfect numbers in the classes already known? Are there any multiperfect numbers beyond the octoperfect class? Are there any odd perfect or multiperfect numbers?

PRIME NUMBERS

Prime numbers, as already mentioned, are those integers which have only one aliquot divisor: 1. The principal method used to determine the primality of a given number, N, is to divide it by all the primes less than $\sqrt{N}$: if a number cannot be divided by 11 (a prime) it is also not divisible by 22 or 33 or 44, or any other multiple of 11. Also, it is not necessary to try primes greater than $\sqrt{N}$ since they would have been discovered during the division by primes smaller than $\sqrt{N}$. This division test is deceptively simple in appearance, but it becomes rather tedious as N gets larger. It would be necessary to divide 18,793 by each of the 33 primes less than $\sqrt{18,793}$ in order to determine the primality of 18,793, for example. Fortunately, electronic computers do such tedious work with incredible rapidity, and the range of prime-number tables is continually being extended. Since there is an infinite number of primes, computers can never complete the task of finding and listing all of them.

The question might be asked: how many primes are there less than some given number, N. By actual count there are 664,579 primes less than 10,000,000, and 5,761,455 primes less than 100,000,000. There are no tables of primes up to 10^{100}, so no count can be made. The French mathematician Adrian M. Legendre (1752–1833) devised a method for determining the exact number of primes less than a given number. The method, though accurate, is extremely cumbersome. For instance, it is calculated that there are 50,847,534 primes less than 1,000,000,000. The procedure for determining this goes as follows: all the multiples of 2 (500,000,000 of them) are subtracted from 1,000,000,000; all the multiples of 3 (333,333,333 of them) are subtracted; then all the multiples of 5, and so on. Corrections must be made: some of the multiples of 3 are also multiples of 2, some of the multiples of 5 are multiples of 2 and 3, and so on. If the reader tries this method to find the number of primes less than 1,000 he will soon appreciate the magnitude of the problem. To determine the number of primes less than 10^{100} would tax the abilities of the fastest known computer.

There is a quick way of determining the *approximate* number of primes less than a given number. It is known as the *prime-number theorem* and states that P, the number of primes less than N, approaches $\frac{N}{\log_e N}$ as N becomes larger. With the aid of a table of logarithms, it becomes a simple matter to determine the approximate, but not the *exact,* number of primes less than a given number, N. The percentage of error decreases as N increases. To take an example, the formula yields an estimate of 48,260,000 primes up to 1,000,000,000, whereas the true count is 50,847,534. The number of primes less than 10^{100} is calculated to be about $(4.34)(10^{97})$, using the prime-number theorem.

There are other formulas that estimate the number of primes less than a given number, but these become more complicated as they become more accurate. What is needed is a formula as simple as that of the prime-number theorem and as accurate as Legendre's enumeration process.

Primes that look "interesting" are a favorite with some mathematicians. Those consisting entirely of ones, the *repunit* primes, can be expressed as $\frac{10^n - 1}{9}$, where n also gives the number of ones. For $n = 7$, the formula gives 1,111,111. If such a repunit number is to be prime, then n must be prime, but the repunit number is then not *necessarily* prime. To date, many possible values for n have been studied. Only 4 of these have yielded prime numbers:

For $n = 2$, the prime is 11
For $n = 19$, the prime is 1,111,111,111,111,111,111
For $n = 23$, the prime is 11,111,111,111,111,111,111,111
For $n = 317$, the prime consists of 317 1's

J. Brillhart and H. C. Williams proved this most recent result ($n = 317$) in 1977. Except for $n = 2$, 19, and 23, there are no smaller repunit primes. Undoubtedly, there *are* larger repunit primes.

A number of laws of prime numbers are known. One of these states that there is at least 1 prime between N and $2N$, where N is greater than 1. A consequence of this law is that there must be at least 3 primes consisting of any number of digits, say 100. Interestingly, 3 such 100-digit numbers were discovered in 1958—though it is estimated that there must be about $(4)(10^{97})$ still awaiting discovery. The full evaluation of these 3 primes has never been published, to the best of my knowledge, and T. Charles Jones, a student at Davidson College in North Carolina, programmed a computer to evaluate them:

$(81)(2^{324}) + 1 =$ 2,768,239,198,553,499,466,783,948,131,195,739, 540,525,105,060,974,221,989,851,365,923,737, 596,764,611,710,152,829,758,864,669,802,497

$(63)(2^{326}) + 1 =$ 8,612,299,728,833,109,452,216,727,519,275,634, 126,078,104,634,142,023,968,426,471,762,739, 189,934,347,542,697,692,583,134,528,274,433

$(35)(2^{327}) + 1 =$ 9,569,221,920,925,677,169,129,697,243,639,593, 473,420,116,260,157,804,409,362,746,403,043, 544,371,497,269,664,102,870,149,475,860,481

Several hundred primes are given by the formula $k2^n + 1$, which produces primes for certain values of k and n. Indeed, the enumeration of the 100-digit primes just listed resulted from studies involving this formula. Extension of these studies should eventually yield a 1,000-digit prime; perhaps a 1,000,000-digit prime might come to light. Such a large number would involve values of n of the order 3,300,000, which could only be reduced appreciably by using extremely large values for k. For example, if n is to be reduced to about 100,000, then k would have to be of the order of $10^{970,000}$.

Another type of prime number, the Mersenne primes, has been mentioned in the discussion of perfect numbers. Much work has been done with Mersenne numbers and primes, but the determination of the primality of the larger Mersenne numbers requires the use of high-speed computers not now available. To determine the next Mersenne prime requires the testing of each prime, p, greater than 11,213 to see if $2^p - 1$ is prime. There is, at present, no way of even determining which of these primes may be *likely* candidates. The next Mersenne prime, if it exists, may be found just a few values beyond 11,213, or several thousand values beyond 11,213.

Prime numbers continue to offer challenges to amateur and professional mathematicians, and a few of the many interesting, but unsolved, problems involving primes will be stated here. Such problems are either technically or theoretically unsolved. The determination of the next Mersenne prime, if it exists, is technically unsolved. "If it exists" suggests, rightly, that the possible existence of another such number is also theoretically unsolved. It is reasonably certain that there are more Mersenne primes and probably more labor will be expended in the physical search than in a theoretical proof of their existence.

Are there arithmetical progressions of primes containing any given number of terms? The sequence of primes 199, 409, 619, 829, 1,039, 1,249, 1,459, 1,669, 1,879, and 2,089 has 10 terms with a common difference of 210. The longest arithmetic progression of primes known, containing but 16 terms, begins with 2,236,133,941, ends with 5,582,526,991, and has a common difference between terms of 223,092,870. Some theoretical work has been done along these lines. To obtain an arithmetic progression or sequence of primes with about 30 terms, it would be necessary to search through the primes in the region of 187,000,000,000 and look for a common difference of approximately 6,000,000,000. Construction of a table of primes covering this region, and the search for a possible 30-term progression in the table, remain technical problems of extremely high order. The possible existence of such a lengthy series of primes remains a theoretical problem. After all, there may not be such a 30-term sequence of primes.

For what values of n does $n^n + 1$ produce primes? If $n = 1, 2$, or 4, the primes 2, 5, and 257 result. No others are known. If $n = 15$, the number has 18 digits, already beyond existing prime tables. Interestingly, though, it has been shown that *if* there are any other primes of this form, they must consist of more than 300,000 digits. Similarly, it has been shown that if primes exist of the form $n^{n^n} + 1$, besides those for $n = 1$ or 2, then such primes must have at least 1,000,000,000,000,000,000 digits.

The formulation of prime-number problems is considerably easier than their solution. If prime numbers are found to have *some* degree of regularity or predictability, perhaps progress could be made. Perhaps, too, they might lose some of their fascination.

AMICABLE NUMBERS

There are only about 1,200 amicable pairs of numbers known, of which some are: 220 and 284 (the smallest pair), 2,620 and 2,924; 6,232 and 6,368; 10,744 and 10,856; 9,363,584 and 9,437,056; and 111,448,537,712 and 118,853,793,424. The largest such numbers known contain 152 digits.

Several methods are available for finding amicable pairs. One common method is the following. Let

$$a = (3)(2^x) - 1$$
$$b = (3)(2^{x-1}) - 1$$
$$c = (9)(2^{2x-1}) - 1$$

If x is greater than 1, and a, b, and c are all primes, then $2^x ab$ and $2^x c$ constitute an amicable pair of numbers. For example, if $x = 4$, then $a = 47$, $b = 23$, and $c = 1{,}151$, which are all primes. Then $(2^4)(47)(23) = 17{,}296$ and $(2^4)(1{,}151) = 18{,}416$. The reader can establish for himself that 17,296 and 18,416 are amicable numbers.

There are, besides amicable pairs, numbers that are amicable triplets. These can be represented as $s(N) = N_1 + N_2$, where $s(N_1) = N + N_2$ and $s(N_2) = N + N_1$. That is, an amicable triplet consists of 3 numbers so related that the sum of the aliquot divisors of any one of them is equal to the sum of the other 2 numbers. One such amicable triplet is: 103,340,640; 123,228,768; and 124,015,008. Another amicable triplet is: 1,945,330,728,960; 2,324,196,638,720; and 2,615,631,953,920. These are not easy to find. The numbers in this last set have 959, 959, and 479 divisors, respectively.

0123456789 OR DIGITAL DIVERSIONS

The 10 digits of the decimal system offer an endless source of mathematical fun. The general idea is to use all digits, although 0 is often excluded, once only to achieve some desired result. Only the usual arithmetical signs and symbols are usually permitted. However, for some particular problems, some of these symbols are not permitted, for other problems many other symbols may be allowed. As we know, the factorial symbol is !. The Greek letter Σ (sigma) indicates successive addition of the integers starting with 1: $\Sigma n = 1 + 2 + 3 + 4 + \cdots + n$. For purposes of reference the following short table is given:

n	$n!$	Σn
0	1	0
1	1	1
2	2	3
3	6	6
4	24	10
5	120	15
6	720	21
7	5,040	28
8	40,320	36
9	362,880	45

A favorite problem is to use the 9 or 10 digits in equations equal to 100. This is fairly easy:

$$(7-5)^2+96+8-4-3-1=100$$
$$3^2+91+7+8-6-5-4=100$$
$$\sqrt{9}-6+72-(1)(3!)-8+45=100$$

If the conditions are restricted to using the digits *in order,* the problem becomes more difficult, but much more interesting. Nevertheless, working in this way, I was able to produce 120 such equations while confining myself to the common arithmetical symbols. The use of just 1 or 2 additional symbols indicated that many, many more equations could be developed. Some of the more interesting results are given below. The first 12 equations are all the possible ones using the 9 digits in order and restricting the use of symbols to the plus and minus signs only. The first equation is the shortest (uses the least number of signs) and the twelfth is the longest.

1. $123-45-67+89=100$
2. $123+4-5+67-89=100$
3. $123+45-67+8-9=100$
4. $1+2+34-5+67-8+9=100$
5. $1+23-4+5+6+78-9=100$
6. $1+23-4+56+7+8+9=100$
7. $12-3-4+5-6+7+89=100$
8. $12+3+4+5-6-7+89=100$
9. $12+3-4+5+67+8+9=100$
10. $123-4-5-6-7+8-9=100$
11. $1+2+3-4+5+6+78+9=100$
12. $-1+2-3+4+5+6+78+9=100$
13. $(1)(2)(3)(4)+5+6+(7)(8)+9=100$
14. $1+(23)(4)-5+6+7+8-9=100$
15. $1+(23)(4)+5-6+7-8+9=100$
16. $(1+2-3-4)(5-6-7-8-9)=100$

17. $[(1)(2)][3! - 4][5][-67 + (8)(9)] = 100$
18. $12 + 3.4 + 5.6 + 7 + (8)(9) = 100$
19. $12(3.4 + 5.6) - 7 + 8 - 9 = 100$
20. $1^{23} + 4 + 5 - 6 + 7 + 89 = 100$
21. $1^{2,345} + 6(7 + 8) + 9 = 100$
22. $1 - (2)(3) + 4 + \sqrt{5^6} - 7 - 8 - 9 = 100$
23. $(1!)(2!)(3! + 4!) - 5! + 6! - [(7!)(8!) \div 9!] = 100$
24. $-\Sigma 1 + (\Sigma 2)(\Sigma 3) + \Sigma 4 + \Sigma 5 + \Sigma 6 + \Sigma 7 - \Sigma 8 + \Sigma 9 = 100$
25. $\sqrt{[(12 \div 3)(\Sigma 4)(5!)] + 6! + \{[(7!)(8)] \div 9\}} = 100$

It is encouraging to know that there appears to be no limit to the range of values of such equations; for example:

$$-(1234) \div (5 - 6 - 7 + 8)(9) = -\infty$$
$$(1234) \times (5 - 6 - 7 + 8)(9) = 0$$
$$(1234) \div (5 - 6 - 7 + 8)(9) = \infty$$

Not much work has been done with the problem in which the digits are used in reverse order, but here are a few such equations:

1. $98 - 76 + 54 + 3 + 21 = 100$
2. $9 - 8 + 76 + 54 - 32 + 1 = 100$
3. $9 - 8 + 7 + 65 - 4 + 32 - 1 = 100$
4. $9 - 8 + 76 - 5 + 4 + 3 + 21 = 100$
5. $98 - 7 - 6 - 5 - 4 + 3 + 21 = 100$
6. $9 + 8 + 76 + 5 + 4 - 3 + 2 - 1 = 100$
7. $9 + 8 + 76 + 5 - 4 + 3 + 2 + 1 = 100$
8. $98 + 7 + 6 - 5 - 4 - 3 + 2 - 1 = 100$
9. $98 + 7 - 6 + 5 - 4 - 3 + 2 + 1 = 100$
10. $98 + 7 - 6 + 5 - 4 + 3 - 2 - 1 = 100$
11. $98 + 7 - 6 - 5 + 4 + 3 - 2 + 1 = 100$
12. $98 - 7 + 6 + 5 + 4 - 3 - 2 - 1 = 100$
13. $98 - 7 + 6 + 5 - 4 + 3 - 2 + 1 = 100$
14. $98 - 7 + 6 - 5 + 4 + 3 + 2 - 1 = 100$
15. $98 - 7 - 6 + 5 + 4 + 3 + 2 + 1 = 100$

16. $-9 + 8 + 7 + 65 - 4 + 32 + 1 = 100$
17. $-9 + 8 + 76 + 5 - 4 + 3 + 21 = 100$
18. $-9 - 8 + 76 - 5 + 43 + 2 + 1 = 100$
19. $(\sqrt{9})(8) + 76 + \sqrt{5 + 4} - 3! + 2 + 1 = 100$
20. $-\sqrt{9} - 8 + 7 + (6! - 5^4 + 3^2)(1) = 100$

The first 18 are the only ones possible when only plus and minus signs are used. The first is the shortest; 8 through 15 are the longest, all having 8 terms; 16, 17, and 18 are the only ones that start with -9 and that use only plus and minus signs.

To continue with other digital diversions, Dudeney once pointed out that the smallest square number containing 9 different digits (excluding 0) is 139,854,276 ($11{,}826^2$) and that the largest such number is 923,187,456 ($30{,}384^2$). Are there square numbers in which the digits are repeated 2, 3, or more times?

There are indeed, and Harry L. Nelson of the Lawrence Radiation Laboratory at the University of California programmed an IBM 7030 computer to solve the problem, yielding the following results in 103 seconds:

1. The smallest and largest 10-digit cases:
   1,026,753,849 ($32{,}043^2$), and
   9,814,072,356 ($99{,}066^2$).
2. The smallest and largest 18-digit cases:
   112,345,723,568,978,496 ($335{,}180{,}136^2$), and
   998,781,235,573,146,624 ($999{,}390{,}432^2$).
3. The smallest and largest 27-digit cases:
   111,222,338,559,598,866,946,777,344 ($10{,}546{,}200{,}195{,}312^2$), and
   999,888,767,225,363,175,346,145,124 ($31{,}621{,}017{,}808{,}182^2$).

The following result is a slight variation on this theme; here each of the 9 digits (excluding 0) appears in each factor:

$$(246{,}913{,}578)(987{,}654{,}312) = (493{,}827{,}156)(493{,}827{,}156)$$

Are there other such results?

Another problem: what 2 or more factors, together containing the 9 (or 10) digits once only, yield a product containing the 9 (or 10) digits once only? My own efforts failed to yield a solution. This problem was submitted to Nelson, and a computer again readily solved the problem. Only the 9-digit case was programmed, for reasons that will soon become apparent. .

The first step toward solution was to list the different types of set-ups that might yield solutions. The following tabulations show them and the number of solutions found for each:

Number of Factors	*Problem Type*	*Number of Solutions*
2	(A)(BCDEFGHI)	567
2	(AB)(CDEFGHI)	342
2	(ABC)(DEFGHI)	370
2	(ABCD)(EFGHI)	346
3	(A)(B)(CDEFGHI)	87
3	(A)(BC)(DEFGHI)	210
3	(A)(BCD)(EFGHI)	244
3	(A)(BCDE)(FGHI)	109
3	(AB)(CD)(EFGHI)	0
3	(AB)(CDE)(FGHI)	52
3	(ABC)(DEF)(GHI)	39
4	(A)(B)(C)(DEFGHI)	16
4	(A)(B)(CD)(EFGHI)	55
4	(A)(B)(CDE)(FGHI)	53
4	(A)(BC)(DE)(FGHI)	59
4	(A)(BC)(DEF)(GHI)	66
5	(A)(B)(C)(D)(EFGHI)	0
5	(A)(B)(C)(DE)(FGHI)	3
5	(A)(B)(C)(DEF)(GHI)	1
5	(A)(B)(CD)(EF)(GHI)	5
6	No Solution	
7	No Solution	
8	No Solution	
9	No Solution	

All 2,624 solutions will not be listed here, but the smallest and largest products for each of the types with two to four factors are tabulated below:

Problem Type	*Smallest Solution*	*Largest Solution*
(A)(BCDEFGHI)	(3)(41,579,862) = 124,739,586	(9)(87,325,146) = 785,926,314
(AB)(CDEFGHI)	(48)(2,573,916) = 123,547,968	(96)(8,745,231) = 839,542,176
(ABC)(DEFGHI)	(627)(198,354) = 124,367,958	(852)(964,713) = 821,935,476
(ABCD)(EFGHI)	(2,598)(47,613) = 123,698,574	(8,352)(97,641) = 815,497,632
(A)(B)(CDEFGHI)	(6)(9)(2,347,851) = 126,783,954	(7)(9)(8,563,124) = 539,476,812
(A)(BC)(DEFGHI)	(9)(83)(165,274) = 123,459,678	(9)(76)(825,143) = 564,397,812
(A)(BCD)(EFGHI)	(9)(162)(84,753) = 123,569,874	(9)(863)(75,142) = 583,627,914
(A)(BCDE)(FGHI)	(9)(2,683)(5,174) = 124,936,578	(9)(7,216)(8,453) = 548,971,632
(AB)(CDE)(FGHI)	(54)(238)(9,617) = 123,597,684	(86)(751)(9,234) = 596,387,124
(ABC)(DEF)(GHI)	(163)(827)(945) = 127,386,945	(531)(876)(942) = 438,176,952
(A)(B)(C)(DEFGHI)	(3)(7)(9)(654,812) = 123,759,468	(6)(8)(9)(751,324) = 324,571,968
(A)(B)(CD)(EFGHI)	(4)(9)(61)(57,283) = 125,793,468	(8)(9)(71)(64,523) = 329,841,576
(A)(B)(CDE)(FGHI)	(7)(9)(516)(3,842) = 124,895,736	(6)(9)(743)(8,521) = 341,879,562
(A)(BC)(DE)(FGHI)	(8)(39)(52)(7,641) = 123,967,584	(7)(64)(81)(9,532) = 345,897,216
(A)(BC)(DEF)(GHI)	(9)(34)(562)(718) = 123,475,896	(8)(92)(531)(746) = 291,548,736

All the solutions to the 5-factor types are shown below:

(A)(B)(C)(DE)(FGHI)	(4)(6)(8)(93)(7,251) = 129,473,856
	(6)(8)(9)(71)(4,523) = 138,729,456
	(6)(7)(9)(51)(8,423) = 162,378,594
(A)(B)(C)(DEF)(GHI)	(6)(7)(9)(524)(831) = 164,597,832
(A)(B)(CD)(EF)(GHI)	(6)(7)(52)(84)(913) = 167,495,328
	(6)(7)(54)(91)(823) = 169,857,324
	(6)(7)(84)(92)(531) = 172,349,856
	(6)(8)(53)(92)(741) = 173,429,568
	(7)(9)(54)(83)(621) = 175,349,286

The smallest product in the complete table of 2,624 solutions is 123,459,678, which is shown in the table above. The largest product is 839,542,176, which is also listed. A number of products result from different sets of factors:

124,367,958 = (627)(198,354) = (9)(26)(531,487)
518,423,976 = (567)(914,328) = (918)(564,732)
172,349,856 = (3)(672)(85,491) = (6)(7)(84)(92)(531)
= (8)(92)(413)(567)
169,857,324 = (36)(4,718,259) = (3,276)(51,849) = (4)(91)(567)(823)
= (6)(7)(54)(91)(823)

There are also solutions using each of the 10 distinct digits. On the basis of some work on the 9-digit solutions I estimate that there are probably 25,000 or more 10-digit answers to this problem. No computer program was set up to determine them, but a total of 6,514 solutions to the 10-digit case can be derived directly from the 2,624 solutions to the 9-digit one by merely appending a 0 to one of the factors and to the product. For example, (3)(41,579,862) = 124,739,586 yields 2 solutions to the 10-digit case:

(30)(41,579,862) = 1,247,395,860
and (3)(415,798,620) = 1,247,395,860

Computer or no computer, the problem could not be left alone. There must be many solutions in which the 0 is in an internal position. I assumed that such a situation might be very similar to a given 9-digit solution, and it was a pleasant surprise to find that it proved true. The 567 two-factor solutions of the type $(A)(BCDEFGHI)$ were the basis for about 1,000 additional 10-digit solutions with internal 0's, such as:

(3)(41,579,862) = 124,739,586 yields (3)(401,579,862) = 1,204,739,586
(6)(27,548,913) = 165,293,478 yields (6)(275,489,013) = 1,672,934,078
(9)(25,763,841) = 231,874,569 yields (9)(257,638,401) = 2,318,745,609

It remains now for some magnanimous company to permit one of its computers to be programmed to determine all the 10-digit solutions to this problem.

NARCISSISTIC NUMBERS

"There are just four numbers, after unity, which are the sums of the cubes of their digits: $153 = 1^3 + 5^3 + 3^3$, $370 = 3^3 + 7^3 + 0^3$, $371 = 3^3 + 7^3 + 1^3$, and $407 = 4^3 + 0^3 + 7^3$. These are odd facts, very suitable for puzzle columns and likely to amuse amateurs, but there is nothing in them which appeals to the mathematician."

So wrote the English mathematician Godfrey H. Hardy (1877–1947) in his book *A Mathematician's Apology*. This section will, I hope, dispel the illusion that such numbers have no appeal to mathematicians, professional or amateur.

Narcissus, according to Greek mythology, fell in love with his own image, seen in a pool of water, and changed into the flower now called by his name. Since this section deals with numbers "in love with themselves" *narcissistic numbers* will be defined as those that are representable, in some way, by mathematically manipulating the digits of the numbers themselves. Each type of manipulation will be dealt with separately in this section.

DIGITAL INVARIANTS

Those integers (>1) that are equal to the sums of the nth powers of their digits are called *perfect digital invariants* (PDI's). The 4 numbers mentioned by Hardy are PDI's of the third order, that is, $n = 3$.

Some numbers are digital invariants in another way. The sum of the nth powers of the digits of a number, N, may equal another number, N_1. The sum of the nth powers of N_1 may equal another number, N_2. If, after repeating this procedure a number of times, the original number, N, is produced, then N is called a *recurring digital invariant* (RDI). Whenever such a series is discovered, the term with the lowest value is called the RDI.

There are 4 such RDI's of the third order: 55, 136, 160, and 919. The recurring sequence for 55 is:

$$
\begin{aligned}
55&: 5^3 + 5^3 = 250 \\
250&: 2^3 + 5^3 + 0^3 = 133 \\
133&: 1^3 + 3^3 + 3^3 = 55
\end{aligned}
$$

A short form for this procedure can be used: 55 → 250 → 133 → 55. The other third-order RDI's and their recurring sequences are:

136 → 244 → 136
160 → 217 → 352 → 160
919 → 1,459 → 919

A PDI of the fourth order is 1,634: $1^4 + 6^4 + 3^4 + 4^4 = 1{,}634$. A fourth-order RDI is 1,138: 1,138 → 4,179 → 9,219 → 13,139 → 6,725 → 4,338 → 4,514 → 1,138.

Many digital invariants of higher orders have been found, but there must be many more awaiting discovery—a challenge to enthusiastic amateurs. It is a marvel that a tenth-order PDI should have been discovered: $4{,}679{,}307{,}774 = 4^{10} + 6^{10} + 7^{10} + 9^{10} + 3^{10} + 0^{10} + 7^{10} + 7^{10} + 7^{10} + 4^{10}$. There must be many other such digital invariants of higher order, but the numbers get larger and more difficult to work with. Recently the work has been extended to the seventeenth order.

In the table below there are some pairs of what may be called ami-

Order	*PDI*	*RDI*
2	None Exist	4
3	153; 370; 371; 407	55; 136; 160; 919
4	1,634; 8,208; 9,474	1,138; 2,178
5	4,150; 4,151; 54,748; 92,727; 93,084; 194,979	9,044; 9,045; 24,584; 58,618
6	548,834	239,459; 282,595
7	1,741,725; 4,210,818; 9,800,817; 9,926,315; 14,459,929	80,441; 86,874; nine more
8	24,678,050; 24,678,051; 88,593,477	6,822; 7,973,187; 8,616,804
9	146,511,208; 472,335,975; 534,494,836; 912,985,153	322,219; 2,274,831; 20,700,388; eleven more
10	4,679,307,774	20,818,070; five more

All solutions for *PDI*'s for the sixth and higher orders are from: Harry L. Nelson, "*More on PDI's*," Publication UCRL–7614, University of California, December 1, 1963.

cable digital invariants (ADI): for order 3 note that $136 \to 244$ and $244 \to 136$ without any intermediate steps; for order 4 there is $2{,}178 \to 6{,}514$ and $6{,}514 \to 2{,}178$; similarly, for order 5 there is $58{,}618 \to 76{,}438$ and $76{,}438 \to 58{,}618$.

Other questions that invite research on a more recreational level are: What connection, if any, is there between the order and the number of invariants in that order, and, are there any PDI's that are also primes?

VR NUMBERS

The VR number 371 has the *V*isible *R*epresentation $3^3 + 7^3 + 1^3$. This has already been termed a PDI. But consider a 6-digit analogue of the above: $165{,}033 = 16^3 + 50^3 + 33^3$. To call this a perfect *digital* invariant is obviously confusing, though it is, in fact, one in the scale of 100. Many similar examples exist in different orders, though much work needs to be done beyond the third order. Almost all of the work following is the result of researches by J. A. Lindon.

1. *Sum of 2 Squares.* Each number is equal to the sum of the squares of the 2 "halves" into which it is divided, such as $12\ 33 = 1{,}233 = 12^2 + 33^2$; $588\ 2{,}353 = 5{,}882{,}353 = 588^2 + 2{,}353^2$. Other numbers include:

12 33	588 2353	17650 38125
88 33	9412 2353	82350 38125
10 100	990 09901	25840 43776
990 100	99010 09901	74160 43776
100 010000	123288 328768	116788 321168
999900 010000	876712 328768	883212 321168

It will be noted that these solutions go in pairs. If $|x||10^n + y|$ is a solution of the $2n$-digit case, so is $|10^{2n} - x||10^n + y|$. For example, $1{,}233 = |12||33|$, $x = 12$, $10^n + y = 33$ and $n = 1$, $y = 23$, therefore $10^{2n} - x = 100 - 12 = 88$, which gives the other solution $|88||33| = 8{,}833$.

No great difficulty would be involved in extending the list. Also, the solution 5,882,353 is a prime, and there may well be many other answers of the same type.

2. *Difference of 2 Squares.* Each number is equal to the square of its right-hand part minus the square of its left-hand part. 266 1,653 = 2,661,653 = $1{,}653^2 - 266^2$. Other numbers include:

4 8	140 400	484 848
	190 476	530 901
34 68	216 513	234 1548
10 101	300 625	266 1653
16 128	334 668	1660 4400
34 188	416 768	3334 6668
		5980 9776

3. *Sum of 2 Triangular Numbers.* Triangular numbers are of the form $\frac{n^2 + n}{2}$, where n is any positive integer. These are not difficult to work out. Ignoring trivial solutions, a satisfactory result in the 6-digit range is readily found: 585 910 = 585,910 = $\Delta_{585} + \Delta_{910}$.

$$\Delta_{585} \text{ means } \frac{585^2 + 585}{2} = 171{,}405 \qquad \Delta_{910} = \frac{910^2 + 910}{2} = 414{,}505$$

and 171,405 + 414,505 = 585,910 = 585 910. Other numbers include:

90 415	700 3676	3030 7171	7804 9756
120 1545	769 3846	4774 8526	8274 9850
150 1726	1474 5226	6154 9231	8455 9880
244 2196	2829 6970	6324 9300	

4. *Sum of 3 Cubes.* Each number is equal to the sum of the cubes of its 3 sections. Thus, 22 18 59 = 221,859 = $22^3 + 18^3 + 59^3$. Other numbers include:

4 18 33	34 00 67	44 46 64	166 500 333
16 50 33	34 10 67	48 72 15	333 667 000
22 18 59	40 70 00	98 28 27	333 667 001
33 67 00	40 70 01	98 32 21	334 000 667
33 67 01			

5. *Amicable Pairs.*

$$3{,}869 = 62^2 + 05^2 \quad \text{and} \quad 6{,}205 = 38^2 + 69^2$$
$$5{,}965 = 77^2 + 06^2 \quad \text{and} \quad 7{,}706 = 59^2 + 65^2$$

6. *Sum of Factorials.*

$$1 = 1!$$
$$2 = 2!$$
$$145 = 1! + 4! + 5!$$
$$40{,}585 = 4! + 0! + 5! + 8! + 5!$$

Note: $0! = 1$.

The first 3 solutions are well known, but the last example was found by Leigh Janes in 1964. Later, Ron S. Dougherty, a student at Davidson College, North Carolina, made an exhaustive search with a computer and found that the above factorial solutions are the only ones that exist.

7. *Sum of Subfactorials.* Factorials have been defined previously in Chapter 2. Subfactorials ($!n$) are not so well-known and they are defined as follows:

$$!n = n!\left[1 - \frac{1}{1!} + \frac{1}{2!} - \frac{1}{3!} + \cdots + (-1)^n\frac{1}{n!}\right]$$

The subfactorials of the digits are: $!0 = 0$, $!1 = 0$, $!2 = 1$, $!3 = 2$, $!4 = 9$, $!5 = 44$, $!6 = 265$, $!7 = 1{,}854$, $!8 = 14{,}833$, $!9 = 133{,}496$.

After his work on factorial solutions, Dougherty was asked to find similar results for subfactorials. There is only one solution to this problem:

$$148{,}349 = !1 + !4 + !8 + !3 + !4 + !9$$

8. *Sums and Powers.* Where a digital invariant was defined as a number equal to the sum of the *n*th powers of its digits, this category of number recreation deals with numbers that are equal to the *n*th power of the sums of their digits, such as:

$$81 = (8 + 1)^2 = 9^2$$
$$4{,}913 = (4 + 9 + 1 + 3)^3 = 17^3$$
$$34{,}012{,}224 = (3 + 4 + 0 + 1 + 2 + 2 + 2 + 4)^6 = 18^6$$

N	$= P^n$
81	9^2
512	8^3
4,913	17^3
5,832	18^3
17,576	26^3
19,683	27^3
2,401	7^4
234,256	22^4
390,625	25^4
614,656	28^4
1,679,616	36^4
17,210,368	28^5
52,521,875	35^5
60,466,176	36^5
205,962,976	46^5

N	$= P^n$
34,012,224	18^6
8,303,765,625	45^6
24,794,911,296	54^6
68,719,476,736	64^6
612,220,032	18^7
10,460,353,203	27^7
27,512,614,111	31^7
52,523,350,144	34^7
271,818,611,107	43^7
1,174,711,139,837	53^7
2,207,984,167,552	58^7
6,722,988,818,432	68^7
20,047,612,231,936	46^8
72,301,961,339,136	54^8
248,155,780,267,521	63^8

$$20{,}864{,}448{,}472{,}975{,}628{,}947{,}226{,}005{,}981{,}267{,}194{,}447{,}042{,}584{,}001 = 207^{20}$$

These numbers can be written as:

$$N = abcd\ldots = (a + b + c + d + \cdots)^n = P^n$$

and a few typical examples are tabulated above.

It is worth noting that there is at least one example where the sum of the digits of an integer is equal to the *exponent* to which another number must be raised to equal the integer: $1{,}180{,}591{,}620{,}717{,}411{,}303{,}424 = 2^{70}$. The sum of the digits in 2^{70} equals 70.

How can examples of the types of numbers tabulated above be found? It certainly would not do to have to hunt through tables of powers of numbers in hopes of finding the desired solutions. Even computer time is too valuable to waste in this type of search. There are ways of eliminating many numbers and limiting the search considerably. The methods to be outlined will be applied to the search for all $N = P^4$ examples.

The most powerful tool to limit such a search is the fact that the sum of the digits of every positive integer gives the same remainder when divided by 9 as does the integer itself when divided by 9. The nth power of the sum of these digits also gives the same remainder when divided by 9. The *digital root* is the ultimate sum of the digits of the integer; for example, the digital root of 496 is 1: $4 + 9 + 6 = 19$, $1 + 9 = 10$, $1 + 0 = 1$. The digital root of any integer is a single digit, so only the remainder of the fourth powers of the nine distinct digits, when divided by 9, need be sought:

d	d^4	*Digital Root of* d^4	*Remainder when Digital Root is Divided by 9*
1	1	1	1
2	16	7	7
3	81	9	0
4	256	4	4
5	625	4	4
6	1,296	9	0
7	2,401	7	7
8	4,096	1	1
9	6,561	9	0

We need only search through tables of P^4 where P yields a remainder of 0, 1, 4, or 7 when divided by 9. This is laborious enough, so the search will have to be limited a bit more.

If N has 10 digits, the maximum sum of these digits is $(9)(10) = 90$, and $90^4 = 65{,}610{,}000$, which is an 8-digit number. Therefore N has less than 9 digits. This information also narrows the search to those P with less than 3 digits. Instead of searching endlessly through tables, it is only necessary to search up to 99 and its fourth power for those values leaving remainders of 0, 1, 4, or 7 when divided by 9. The search can be narrowed even further, however.

If N has 8 digits, then the maximum sum of these digits would be $(8)(9) = 72$, and $72^4 = 26{,}873{,}856$. The number that is less than 26,873,856 and has the maximum sum for its digits is 19,999,999. The sum of these digits is 64 and $64^4 = 16{,}777{,}216$. It is now known that the first digit of N is less than 2. Another fact that can be used is that the fourth powers of integers terminate with 0, 1, 5, or 6. Since the maximum value for the terminal digit of a fourth power is 6, the integer less than 16,777,216 which has the maximum sum for its digits is 9,999,996. The sum of these digits is 60, and $60^4 = 12{,}960{,}000$.

Instead of having to search to 99, the search is limited to integers up to 60. It is not difficult to divide each integer from 1 to 60 by 9 and determine those that leave remainders of 0, 1, 4, or 7. They are: 1, 4, 7, 9, 10, 13, 16, 18, 19, 22, 25, 27, 28, 31, 34, 36, 37, 40, 43, 45, 46, 49, 52, 54, 55, 58. Only the fourth powers of 7, 22, 25, 28, and 36 yield solutions to the problem. There are a few other refined methods that can help to narrow the search even further, but such work may become more complicated than merely searching through the tables.

The above methods were used by T. Charles Jones of Davidson College to determine all such numbers up to and including P^{101}. All values of n in P^n are represented in the range from $n = 2$ to $n = 101$, and there is a total of 432 examples of these numbers in this range. All cannot be given here, but it is interesting to show at least one P^{101} example in full:

$1468^{101} =$ 69076705 04793461 45526875 25633286 29356758 06290369 57264412
76537533 01688875 87647821 39542833 19340073 41871874 06338859
59764845 15235943 07415120 29562743 76018961 33803755 77309227
54848168 47699952 43473825 72197272 51077441 15428499 32284188
41599566 22824107 08524949 37732305 04940602 64582216 28159056
71302903 95809952 85801616 72423705 07603968

The sum of all the digits in this 320-digit number is 1,468, which is equal to P. This is the largest known number of this type.

9. *Miscellaneous*. Related to digital invariants are the following interesting examples:

$$43 = 4^2 + 3^3 \qquad 1{,}306 = 1^1 + 3^2 + 0^3 + 6^4$$
$$63 = 6^2 + 3^3 \qquad 1{,}676 = 1^1 + 6^2 + 7^3 + 6^4$$
$$89 = 8^1 + 9^2 \qquad 2{,}427 = 2^1 + 4^2 + 2^3 + 7^4$$
$$135 = 1^1 + 3^2 + 5^3 \qquad 3{,}435 = 3^3 + 4^4 + 3^3 + 5^5$$
$$175 = 1^1 + 7^2 + 5^3 \qquad 438{,}579{,}088 = 4^4 + 3^3 + 8^8 + 5^5$$
$$518 = 5^1 + 1^2 + 8^3 \qquad + 7^7 + 9^9 + 0^0$$
$$598 = 5^1 + 9^2 + 8^3 \qquad + 8^8 + 8^8$$

The last two examples are the only ones of their type known.

MULTIGRADES

A *multigrade* is a particular relationship between sets of numbers and their powers. Here are three cases:

$$1^n + 6^n + 8^n = 2^n + 4^n + 9^n \qquad (n = 1, 2)$$
$$1^n + 5^n + 8^n + 12^n = 2^n + 3^n + 10^n + 11^n \quad (n = 1, 2, 3)$$
$$1^n + 5^n + 8^n + 12^n + 18^n + 19^n = 2^n + 3^n + 9^n + 13^n + 16^n + 20^n$$
$$(n = 1, 2, 3, 4)$$

Unless a proof is outlined, a conventional shorthand in writing multigrades will be used. The sums of the first powers are indicated by S_1, the sums of the second powers by S_2, and so on. The three examples above can be written in shorthand form as follows, where the number above the equal sign indicates the highest power (or exponent) to which the numbers are to be raised.

$$1, 6, 8 \stackrel{2}{=} 2, 4, 9 \qquad S_1 = 15, S_2 = 101$$
$$1, 5, 8, 12 \stackrel{3}{=} 2, 3, 10, 11 \qquad S_1 = 26, S_2 = 234, S_3 = 2{,}366$$
$$1, 5, 8, 12, 18, 19 \stackrel{4}{=} 2, 3, 9, 13, 16, 20 \qquad S_1 = 63, S_2 = 919,$$
$$S_3 = 15{,}057, S_4 = 260{,}755$$

One property of a multigrade is that the same quantity can be added to each term without affecting the relationship. The first example above can be changed by adding 7 to each term to form:

$$8, 13, 15 \stackrel{2}{=} 9, 11, 16 \qquad S_1 = 36, S_2 = 458$$

Forming a multigrade is easy. Start with a simple equality such as $1 + 4 = 2 + 3$. Now add 4 to each term: $5 + 8 = 6 + 7$.

A second-order multigrade can be obtained by "switching sides" and combining, yielding:

$$1^n + 4^n + 6^n + 7^n = 2^n + 3^n + 5^n + 8^n \qquad (n = 1, 2)$$

Four was added since that increase is the smallest that would result in the second-order multigrade with all the terms different.

A third-order multigrade can be built up by using the newly formed second-order multigrade and one formed by adding 8 to each of its terms:

$$9^n + 12^n + 14^n + 15^n = 10^n + 11^n + 13^n + 16^n \qquad (n = 1, 2)$$

Switching sides in these two equations and combining as before gives us:

$$1^n + 4^n + 6^n + 7^n + 10^n + 11^n + 13^n + 16^n$$
$$= 2^n + 3^n + 5^n + 8^n + 9^n + 12^n + 14^n + 15^n \qquad (n = 1, 2, 3)$$

Here is a fifth-order multigrade derived by another method:

$$1, 9, 18, 38, 47, 55 \stackrel{5}{=} 3, 5, 22, 34, 51, 53$$
$$(S_1 = 168, S_2 = 7{,}084, S_3 = 331{,}632, S_4 = 16{,}226{,}980, S_5 = 813{,}813{,}168)$$

The following fifth-order multigrade is of considerably lower magnitude:

$$1, 6, 7, 17, 18, 23 \stackrel{5}{=} 2, 3, 11, 13, 21, 22$$
$$(S_1 = 72, S_2 = 1{,}228, S_3 = 23{,}472, S_4 = 472{,}036, S_5 = 9{,}770{,}352)$$

The late Royal V. Heath devised a large number of unusual multigrades:

$$12, 43, 65, 78 \stackrel{2}{=} 87, 56, 34, 21 \qquad (S_1 = 198, S_2 = 12{,}302)$$

Note that the digits from 1 to 8 are used and that the two sides are reversed.

Another quite unusual Heath multigrade:

$$1{,}118;\ 1{,}881;\ 8{,}181;\ 8{,}818 \stackrel{2}{=} 1{,}181;\ 1{,}818;\ 8{,}118;\ 8{,}881$$
$$(S_1 = 19{,}998;\ S_2 = 149{,}473{,}970)$$

If this multigrade is reversed, another is formed:

$$1{,}888;\ 8{,}118;\ 8{,}181;\ 1{,}811 \stackrel{2}{=} 8{,}188;\ 1{,}818;\ 1{,}881;\ 8{,}111$$
$$(S_1 = 19{,}998;\ S_2 = 139{,}674{,}950)$$

MISCELLANEOUS

Cubes

The *American Mathematical Monthly* of January 1957 asked for an integer less than 1,000 whose cube could be represented in 5 distinct ways as the sum of the cubes of 3 positive integers. That is, find $N^3 = a^3 + b^3 + c^3$ where (a, b, c) represent 5 distinct sets of values and N is less than 1,000.

In the table below, three solutions to the problem are shown that have gone far beyond the original request. David A. Klarner of Edmonton, Alberta, supplied those in the first two columns, and Leon Bankoff of Los Angeles, California, furnished that in the last column.

$N = 492$			$N = 792$			$N = 870$		
a	b	c	a	b	c	a	b	c
24	204	480	30	456	738	17	687	694
48	85	491	88	528	704	72	486	816
72	384	396	108	184	788	200	540	790
113	264	463	188	298	774	225	630	735
144	360	414	189	387	756	235	485	810
176	204	472	225	279	774	260	550	780
207	297	438	288	414	738	330	450	810
226	332	414	292	540	680	380	480	790
246	328	410	374	429	715	435	580	725
281	322	399	396	528	660	537	564	687
			480	542	610			

FACTORIAL PRODUCTS

What factorials are the products of factorials of numbers in arithmetic sequence or progression? At present only six solutions are known:

$$(0!)(1!) = 1! \qquad (6!)(7!) = 10!$$
$$(1!)(2!) = 2! \qquad (1!)(3!)(5!) = 6!$$
$$(0!)(1!)(2!) = 2! \qquad (1!)(3!)(5!)(7!) = 10!$$

It is easy, by the way, to find any number of examples of integers that are nonconsecutive and not in arithmetic progression, such that the product of their factorials equals another factorial:

$$(4!)(23!) = 24!$$
$$(2!)(4!)(47!) = 48!$$
$$(2!)(3!)(4!)(287!) = 288!$$

PRINTER'S "ERRORS"

Dudeney, in his *Amusements in Mathematics,* shows a printer's error that turned out to be correct: $2^5 9^2 = 2{,}592$. Donald L. Vanderpool, an enthusiastic puzzlist from Towanda, Pennsylvania, worked out a number of similar cases:

$$2^5 \cdot \tfrac{25}{31} = 25\tfrac{25}{31}$$
$$11^2 \cdot 9\tfrac{1}{3} = 1{,}129\tfrac{1}{3}$$
$$21^2 \cdot 4\tfrac{9}{11} = 2{,}124\tfrac{9}{11}$$

Many solutions can be used to generate an infinite number of "errors":

$$13^2 \cdot 7\tfrac{6}{7} = 1{,}327\tfrac{6}{7}$$
$$13^2 \cdot 7{,}857{,}142\tfrac{6}{7} = 1{,}327{,}857{,}142\tfrac{6}{7}$$
$$13^2 \cdot 7{,}857{,}142{,}857{,}142\tfrac{6}{7} = 1{,}327{,}857{,}142{,}857{,}142\tfrac{6}{7}$$
$$\cdots$$

$$3^4 \cdot 425 = 34{,}425$$
$$3^4 \cdot 4{,}250 = 344{,}250$$
$$3^4 \cdot 42{,}500 = 3{,}442{,}500$$
$$\cdots$$

$$31^2 \cdot 325 = 312{,}325$$
$$31^2 \cdot 3{,}250 = 3{,}123{,}250$$
$$\ldots$$

The occurrence of exponents and fractions was a bit annoying so I found the following infinite number of solutions:

$$73 \cdot 9 \cdot 42 = 7 \cdot 3{,}942$$
$$73 \cdot 9 \cdot 420 = 7 \cdot 39{,}420$$
$$\ldots$$

AUTOMORPHIC NUMBERS

An *automorphic number* is one whose square ends with the given number: $5^2 = 25$; $76^2 = 5{,}776$; $625^2 = 390{,}625$. Thus, 5, 76, and 625 are automorphic numbers.

Only two automorphic numbers have any given number of digits, and these numbers are intimately related. Except for the final digits, the sum of the corresponding digits in the pair sums to 9. J. A. H. Hunter found a pair of 17-digit automorphic numbers early in 1964, which was the record for that time. R. A. Fairbairn, of Toronto, Ontario, and a friend of Mr. Hunter, was determined to do a bit better. Eventually, Mr. Fairbairn went overboard and produced a pair of 100-digit automorphic numbers! Here they are, in full:

3,953,007,319,108,169,802,938,509,890,062,166,509,580,863,811,000,557,
423,423,230,896,109,004,106,619,977,392,256,259,918,212,890,625

and

6,046,992,680,891,830,197,061,490,109,937,833,490,419,136,188,999,442,
576,576,769,103,890,995,893,380,022,607,743,740,081,787,109,376

The devoted number enthusiast might check these out by squaring either one of them to see if the product terminates with the original number. More recent computer work has extended the size of automorphic numbers to more than 25,000 digits!

These numbers actually constitute a table of all known automorphic numbers. Dropping digits successively from the left of each of these numbers forms new automorphic ones, though with fewer digits. Excluding those that would begin with a 0, there are 167 different automorphic numbers shown in the 100-digit examples above.

TWO RATHER LARGE NUMBERS

What is the largest number that can be written using *only* three digits? It turns out to be 9^{9^9} which is $9^{387,420,489}$. The number, if calculated and written out at 5 digits to an inch, would take nearly 1,167 miles of paper strip.

This particular number has fascinated mathematicians for over a half a century. Since 1906, and possibly earlier, work has been done to determine as many of its digits as possible. There are 369,693,100 digits in the number and that would almost guarantee that the number never will be fully evaluated. However, the last 2,000 digits and the first 1,200 digits of the number were calculated recently by Fred Gruenberger, a mathematician at the RAND Corporation. To satisfy the curiosity of the reader, I will give only the first and last 10 digits of this fantastic number:

4,281,247,731 . . . 2,627,177,289

Although the last 9 or 10 digits of the first 1,200 digits are in doubt, there is still a total of 3,190 known digits. Even if the remaining spaces were filled in by a random array, the error involved would be infinitesimal, something of the order of 1 part in $10^{1,190}$.

As a testimonial to the lengths that some number enthusiasts will go, I will mention one more large number: $9^{9^{9^9}}$. A calculation shows that this number, if written out in full at 5 digits to an inch, would take about $1.16 \times 10^{369,693,094}$ miles of paper strip. Even if the *ink* used in printing this number was in a layer but 1 atom thick, there would not be enough total matter in millions of our universes to print the number. Nevertheless, with a knowledge of the elementary properties of numbers and a simple desk calculator, the last 10 digits of this fantastically huge number have been calculated: . . . 1,045,865,289.

This chapter has gone rather thoroughly into the purest of mathematical recreations, namely, recreations with numbers only. The next chapter also deals with recreations with numbers, but in a different way. No knowledge of mathematics beyond the elementary operations of addition, subtraction, multiplication, and division is required.

7

.:: alphametics

Cryptarithms are puzzles in which letters or symbols are substituted for the digits in an arithmetical calculation. Algebraic expressions might be regarded as cryptarithms of a sort, but algebra is not generally considered to be mathematically recreational. Cryptarithms have existed for centuries, and it is doubtful if it will ever be known when such puzzles were first devised. If a cryptarithm utilizes letters in place of the digits, and these letters form sensible words or phrases, the puzzle is termed an *alphametic*. J. A. H. Hunter coined the term in 1955.

Examine the simple addition problem shown below:

```
  9,567
  1,085
 ------
 10,652
```

Now substitute a single symbol, such as an asterisk, in place of all the digits:

```
  ****
  ****
 -----
 *****
```

The cryptarithm thus formed could be solved with ease since there are quite a number of 5-digit integers that result from the addition of two 4-digit integers. Only one digit can be solved with certainty: the first asterisk in the sum must be 1, since the sum of 2 digits is never greater than 18, or 19 if there has been a carry-over from the addition of 2 digits in the previous column. The mere existence of a multitude of solutions renders this cryptarithm unsuitable as a puzzle.

A proper cryptarithm should be solvable without an inordinate amount of tedious work and yield a single unique answer. Instead of the single symbol used in the example above, an *X* might be substituted in place of all the 0's, an *R* in place of all the 1's, a *Z* for the 2's, and so on. In this way the following cryptarithm might be formed:

```
  T Q S V
  R X W Q
---------
R X S Q Z
```

Now, although the problem is confusing in appearance, sufficient information is given so that a series of logical deductions and arithmetical work will yield the original addition problem and no other.

The distinction between a cryptarithm and an alphametic in this particular problem lies in the choice of letters making up the appearance of the puzzle. Proper substitution of letters for the digits in the addition problem above can yield:

```
  B E S T
  M A D E
---------
M A S E R
```

This is certainly more interesting than the previous set-up. It imparts no more information than the first letter substitution, but it certainly attracts more attention and invites more puzzlists to solve the problem.

Any number of possible letter substitutions can be made and still yield words and phrases. The following alphametics can be formed from the same problem, either as another addition or as a subtraction problem:

```
  B A R E      C O U N T
  F E L T      - C O I N
---------      ---------
B A L E D        S N U B
```

All cryptarithms and alphametics should be solvable by the application of elementary mathematical knowledge and logic—and a certain amount of honest work. It is remarkable that this form of puzzle, which calls for no more knowledge of mathematics than elementary arithmetic, can offer considerable challenge to graduate mathematics students and professional mathematicians. An extensive mathematical education will not permit one to solve these puzzles "at a glance," but continued practice at solving them will instill a bit of facility in the art.

The general ideas involved in solving alphametics can best be shown by solving a simple problem:

$$(\text{BE})(\text{BE}) = \text{MOB}$$

Here a 3-digit number is the product of a 2-digit number multiplied by itself. Basic knowledge of the laws of multiplication will immediately force the conclusion that B cannot be greater than 3. For if B is 4, and the lowest possible value, 0, is assigned to E then BE = 40. However, (40)(40) = 1,600, a 4-digit number, and the product in the puzzle to be solved has but 3 digits. Convention demands that the initial letters or symbols of alphametics cannot be 0, so B is either 1, 2, or 3. Another convention demands that 2 different letters cannot be substituted for the same digit. That is, if B turns out to be 3, then no other letter in this alphametic could stand for 3. Attention can be directed to E since much can be deduced from the fact that (E)(E) ends in B. If E equals 0, 1, 5, or 6, then the product would be a number ending in 0, 1, 5, or 6, respectively. Since the product, MOB, does not end in E, these numbers for E are eliminated. 2, 3, 4, 7, and 8 can also be eliminated as values for E, since they would yield the terminal digits of 4, 6, or 9 for MOB, and B has been established as being 1, 2, or 3. Only one value for E, 9, remains: (9)(9) = 81, so B = 1, and the alphametic is solved: (BE)(BE) = MOB is (19)(19) = 361.

The above paragraph can be phrased in a shorter form using a few mathematical symbols: $\neq$ means "not equal to"; $<$ means "less than"; $\leqslant$ means "less than or equal to"; $>$ means "greater than"; $\geqslant$ means "greater than or equal to."

$B = 1, 2,$ or 3 since $B \neq 0$ and $B \geqslant 4$ yields a 4-digit product.

$E \neq 0, 1, 5,$ or 6, since B would have to equal E.

$E \neq 2, 3, 4, 7,$ or 8, since B would have to be 4, 6, or 9; but $B = 1$, 2, or 3.

Only $E = 9$ remains and $(9)(9) = 81$, and therefore $B = 1$, $BE = 19$, and $MOB = 361$.

That was a very simple example and it might be expected that work of a more sophisticated nature is required for most alphametics. This will be illustrated by starting the solution of the BEST MADE MASER alphametic.

The rules of addition force the conclusion that $M = 1$. Then, $M + B = 1 + B = 10$, so that $A = 0$ and $B = 8$ or 9. If $B = 8$, then there must be a carry-over of 1 from the addition of $0 + E = S$. This could only be true if $E = 9$, but then S would have to be 0. Since 0 has been assigned to A, then S cannot be 0 and B cannot be 8. Therefore, $B = 9$. Substituting these established values into the alphametic:

$$\begin{array}{r} B\ E\ S\ T \\ M\ A\ D\ E \\ \hline M\ A\ S\ E\ R \end{array} \quad \text{yields} \quad \begin{array}{r} 9\ E\ S\ T \\ 1\ 0\ D\ E \\ \hline 1\ 0\ S\ E\ R \end{array}$$

The second column of addition indicates that $S = E + 1$. E cannot be 8, since this would force S to be 9, and it has been established that B is 9. $S + D$ ends in E and, since $S = E + 1$, it must follow that $D + 1$ ends in 0. D must be either 8 or 9 to yield such a result. Since $B = 9$, then $D = 8$ and the alphametic now appears as:

$$\begin{array}{r} B\ E\ S\ T \\ M\ A\ D\ E \\ \hline M\ A\ S\ E\ R \end{array} \quad \text{yields} \quad \begin{array}{r} 9\ E\ S\ T \\ 1\ 0\ 8\ E \\ \hline 1\ 0\ S\ E\ R \end{array}$$

All the above reasoning reduces the possibilities for E to 2, 3, 4, 5, or 6, and the corresponding values for S to 3, 4, 5, 6, or 7. If $E = 2$ and $S = 3$, then $S + 8 = 12$ (in the third column of addends), which implies a carry-over of 1 from the last column. This could only occur if $T + E$ is greater than 9. If $E = 2$, then $T = 8$ or 9. But these values have already been assigned to D and B, respectively. Therefore E and S cannot be 2 and 3. Similar trials with the other pairs of values for E and S eventually yield E and S as 5 and 6. These nearly complete the alphametic except for T and R, which can then be solved readily enough.

The two conventions previously mentioned—initial letters or symbols cannot be 0, nor can 2 or more different letters or symbols stand for the same digit—may be discarded for some particular puzzle, but then a clue must be supplied indicating this.

Often, some interesting alphametics have more than one solution and clues may be provided to eliminate all but the desired one. (BE)(BE) = ARE has 2 solutions (BE = 16 or 31), but a "clue" that BE is even makes only one solution acceptable. (BE)(BE) = ABE has but one solution (BE = 25) while (BE)(BE) = SAD has 4 distinct solutions (BE = 17, 18, 24, or 29). The "clue" for this last alphametic might be to find the solution minimizing the SADness.

It is interesting to see just how many solutions might result if the unique-solution rule for a "good" alphametic or cryptarithm is temporarily ignored. Dudeney published a number of alphametics during his career as a puzzlist, and one he devised in 1924 is the following subtraction alphametic:

$$\begin{array}{r} \text{E I G H T} \\ -\ \text{F I V E} \\ \hline \text{F O U R} \end{array}$$

There are 10 distinct solutions to this puzzle:

12,348 − 6,291 = 6,057	16,743 − 8,651 = 8,092
12,375 − 6,281 = 6,094	16,725 − 8,631 = 8,094
12,780 − 6,231 = 6,549	16,905 − 8,671 = 8,234
14,820 − 7,461 = 7,359	17,036 − 8,791 = 8,245
15,230 − 7,541 = 7,689	17,054 − 8,761 = 8,293

Since the values for U and V are interchangeable, an additional 10 solutions can be written, making a total of 20. While such a multitude of solutions is not usually desirable in an alphametic, at times it might be good practice to work on such "faulty" puzzles as a test of one's ability to find all the solutions.

A set of criteria for an ideal alphametic will be given here:

1. Only 9 or 10 different letters should be used, representing the 9 or 10 digits in the decimal system. Some leniency would be allowable: an alphametic might contain more than 9 or 10 letters, though not necessarily all different. The fifth alphametic in the collection given later is an example of just such a puzzle.

2. The letters should form words or phrases that make sense.

3. The alphametic should be solvable by logic rather than trial and error.

4. There should be but one solution.

An attempt to use only 9 or 10 different letters in a given alphametic does not guarantee a unique solution. Here are 2 addition alphametics utilizing only 10 different letters, but both alphametics have several solutions:

$$\begin{array}{r} \text{C A N} \\ \underline{\text{S H E}} \\ \text{Q U I T} \end{array} = \begin{array}{r} 324 \\ \underline{765} \\ 1{,}089 \end{array} = \begin{array}{r} 437 \\ \underline{589} \\ 1{,}026 \end{array} \ \ldots$$

$$\begin{array}{r} \text{U} \\ \text{G O} \\ \underline{\text{F A R}} \\ \text{W I D E} \end{array} = \begin{array}{r} 5 \\ 84 \\ \underline{973} \\ 1{,}062 \end{array} = \begin{array}{r} 7 \\ 85 \\ \underline{934} \\ 1{,}026 \end{array} \ \ldots$$

Cryptarithms have even been studied with electronic computers. Charles L. Baker, a mathematician at RAND Corporation, programmed a computer to solve the following series of cryptarithms:

(ABCD)(E) = FGHIJ	(F not 0)
(ABCD)(E) = FGHI	(No 0 at all)
(ABC)(DE) = FGHIJ	(F not 0)
(ABC)(DE) = FGHI	(No 0 at all)

The first and third equations constitute two cryptarithmic multiplication problems in which the 10 digits appear only once each; the second and fourth equations utilize 9 distinct digits. It might appear that each of these equations has a unique solution or, at most, 2 or 3. Actually, there are 13 distinct answers for the first equation, only 2 for the second, 9 for the third, and 7 for the fourth. Examination of the following tabulation will disclose a number of interesting pairs in which the products are equal or have permutations of the same digits.

(ABCD)(E) = FGHIJ
(3,907)(4) = 15,628
(5,694)(3) = 17,082
(6,819)(3) = 20,457
(6,918)(3) = 20,754
(3,094)(7) = 21,658
(8,169)(3) = 24,507
(9,168)(3) = 27,504
(7,039)(4) = 28,156
(4,093)(7) = 28,651
(5,817)(6) = 34,902
(9,127)(4) = 36,508
(9,304)(7) = 65,128
(9,403)(7) = 65,821

(ABCD)(E) = FGHI:
(1,738)(4) = 6,952
(1,963)(4) = 7,852

(ABC)(DE) = FGHIJ:
(402)(39) = 15,678
(297)(54) = 16,038
(594)(27) = 16,038
(495)(36) = 17,820
(396)(45) = 17,820
(367)(52) = 19,084
(345)(78) = 26,910
(715)(46) = 32,890
(927)(63) = 58,401

(ABC)(DE) = FGHI:
(157)(28) = 4,396
(297)(18) = 5,346
(198)(27) = 5,346
(483)(12) = 5,796
(138)(42) = 5,796
(186)(39) = 7,254
(159)(48) = 7,632

There are a number of other types of multiplication cryptarithms utilizing 9 or 10 distinct digits. Some can have no solution: (A)(B) = CDEFGHI or CDEFGHIJ; (A)(BC) = DEFGHI or DEFGHIJ; (AB)(CD) = EFGHI or EFGHIJ; and (A)(B)(C)(D)(E) = FGHI or FGHIJ. Others may have answers, but I know of none that have been discovered: (AB)(CD)(E) = FGHI or FGHIJ; (A)(BC)(DEF) = GHIJ.

Unless otherwise specified, alphametics are assumed to be solvable in the decimal system. However, many such puzzles can be worked in other numerical systems as well. What follows is a remarkable case of multiple solutions when the decimal-system restriction is removed.

The following alphametic was composed by Hunter and published in *Recreational Mathematics Magazine:*

```
  T H E S E
  T E A S E
  T R I E D
-----------
R E A D E R
```

Readers of the magazine were challenged to find a solution to the alphametic if the word TIRED replaced TRIED; few, if any, results were expected. Such underestimation is hard to come by!

There were two oversights in the framing of the problem: no restriction was made concerning the use of a 0 for initial letters, and no restriction was placed on the use of any particular system of numeration.

In the decimal system there is one solution to the TIRED variation, and then only if R is equal to 0; it is given below. If the usual convention, initial letters cannot equal 0, is followed, there are a number of solutions in other systems. These 8 answers were found quite readily by a number of puzzle solvers:

THESE	*TEASE*	*TIRED*	*READER*	
29,747	27,847	21,076	078,670	Base 10
18,454	14,954	17,043	049,340	Base 11
89,353	83,153	80,237	231,732	Base 11
90,656	96,356	97,261	263,162	Base 11
64,757	67,257	63,17*E*	172,*E*71	Base 12
72,*XEX*	7*X*,4*EX*	73,1*X*5	1*X*4,5*X*1	Base 12
78,*E*5*E*	7*E*,25*E*	76,1*E*3	1*E*2,3*E*1	Base 12
7*X*,*E*5*E*	7*E*,25*E*	74,1*E*3	1*E*2,3*E*1	Base 12

H and I are interchangeable; $X = 10$ and $E = 11$ in the base-12 solutions. X and E are single digits in the base-12 system. While these answers are quite interesting, they are also complex, and the neophyte will find it better not to attempt many alphametic solutions in systems other than the decimal. There are no solutions to the above puzzle using systems with bases less than 10.

Lowell Carmony, an alphametics enthusiast and mathematics major at Indiana University, found all 8 solutions, and a post card was dispatched in jest to him requesting a solution in base 119344307. In suspiciously short time Mr. Carmony replied with the formula yielding all 59,672,141 solutions. No more post cards were sent, and it was thought that the end had been reached.

Alas, the 59,672,149 solutions mentioned above turned out to be far short of the ultimate performance. Very soon after Mr. Carmony's astounding work, Mrs. Rae Clair Nelson of Livermore, California, submitted a formula yielding an *infinite* number of solutions. For the edification of the nonbelieving reader the formula is given below.

The solution to the TIRED variation in any base L, where $L = 6k + 4$ $(k = 2, 3, 4, 5, 6 \ldots)$, $\text{H} + \text{I} = L - 6$, and $\text{T} = \dfrac{(2L+1)}{3}$ is:

$$\begin{array}{rrrrrr} & \text{T} & \text{H} & \text{E} & \text{S} & \text{E} \\ & \text{T} & \text{E} & \text{A} & \text{S} & \text{E} \\ & \text{T} & \text{I} & \text{R} & \text{E} & \text{D} \\ \hline \text{R} & \text{E} & \text{A} & \text{D} & \text{E} & \text{R} \end{array} \quad = \quad \begin{array}{rrrrrr} & \text{T} & \text{H} & 1 & L/2 & 1 \\ & \text{T} & 1 & (L-4) & L/2 & 1 \\ & \text{T} & \text{I} & 2 & 1 & 0 \\ \hline 2 & 1 & (L-4) & 0 & 1 & 2 \end{array}$$

For example, a solution in base 28 is:

$$\begin{array}{rrrrrr} & 19 & 17 & 1 & 14 & 1 \\ & 19 & 1 & 24 & 14 & 1 \\ & 19 & 5 & 2 & 1 & 0 \\ \hline 2 & 1 & 24 & 0 & 1 & 2 \end{array}$$

where "14," "17," "19," and "24" are single digits in the base-28 system. By using the indicated formula, solutions can be found very quickly in base 16, 22, 28, 34, 40, ... It might appear that if $k = 1$, then $L = 10$, and a solution might be found in base 10 in which R is *not* 0; try it, though.

Now that the methods of solution and peculiarities of alphametics and cryptarithms have been discussed, a generous supply of them is offered. Their answers appear on pages 194–200.

ALPHAMETIC PUZZLES

1. For a starter, an easy alphametic is given. Each letter always stands for the same digit, but an asterisk may stand for any digit.

```
AN ) EASY ( ONE
     *T
     ---
     ***
     **R
     ----
       **
       *Y
       --
```

2.

```
  BASE
  BALL
 -----
 GAMES
```

3. As might have been expected, France could not be fitted into *this* alphametic. (2 solutions.)

```
    UK
   USA
  USSR
 -----
 ABOMB
```

4. However, France *could* be paired together, with certain modifications, with China.

```
  FRENCH
 - CHINO
 -------
    BOMB
```

5. There is a classic alphametic, SEND + MORE = MONEY, which has been known since 1914. Research into the origin of this classic revealed that it was not an appeal by an impoverished college lad, but, rather, to Dad from his daughter. His reply is noted here. What is the largest amount of CASH he could have sent?

```
 A L A S
 L A S S
     N O
 M O R E
 -------
 C A S H
```

6. Upon receipt of the sad tidings from Dad, the lass transmitted another plea:

S E N D ÷ A = G I F T

Well, the GIFT, *odd* as it was, appeared to satisfy the lass.

7. Alphametic pairs are not easy to come by. In the pair below, an A stands for the same digit in both alphametics. However, neither can be solved uniquely by itself. Both must be solved together.

```
a.      L E T        b.    S E T S
        T H E              T I R E
      -------            ---------
      L A S S            L A T E R
```

The influence of the space age is felt in the next 3 alphametics.

```
8.    T R A C K
      S P A C E
  -------------
    R O C K E T

9.    M O O N
        M E N
        C A N
    ---------
    R E A C H
```

10. This alphametic concerns the first dual orbital flight; it was composed by two persons from two different countries, and it has 2 distinct solutions.

```
    N O W
    T W O
      I N
  T W I N
---------
O R B I T
```

11. Sermonizing is hard to avoid, even in a book of puzzles and mathematical recreations. Again, the asterisks indicate any digit.

```
    S I N
      I N
  -------
  * * * *
  U * S
---------
* E V I L
```

Some of the more interesting alphametics are those that are doubly true: the word sums are as correct as the numerical sums. The next 5 alphametics illustrate this point.

12.

```
    O N E
    T W O
  F I V E
---------
E I G H T
```

13.

```
      T W O
  T H R E E
  S E V E N
-----------
T W E L V E
```

14. This is another true addition, where even TWO and TWENTY are even.

```
      T W O
  S E V E N
E L E V E N
-----------
T W E N T Y
```

15. Obviously 1 + 1 + 2 + 12 = 16. Conveniently, SEIZE is divisible by 16 yielding 2 solutions to this puzzler.

```
      U N
      U N
  D E U X
D O U Z E
---------
S E I Z E
```

16. Occasionally, though, an alphametic is only half-true as is proven by this puzzler, which can be solved and be true by a substitution in another base system. Ignoring the poor addition, though, there is a solution in base 10.

```
    N I N E
  T H R E E
  S E V E N
-----------
T W E L V E
```

17. This is a neat little problem, since it has 2 solutions: one as an addition problem, the other as subtraction.

```
S Q U A R E
  D A N C E
-----------
D A N C E R
```

18. The grammar may be atrocious, but the alphametic is not.

```
      C A T S
        E A T
    ---------
    * * * * *
  * * * * *
* * * * *
-------------
* M O U S E S
```

19. A very appropriate puzzle for April 1.

```
A L L ) F O O L S ( D A Y
        G * *
        -----
        * A * *
        * * M *
        -------
          * * E *
          * * * S
          -------
```

20. EVE double talked. There are 2 solutions here.

$$\frac{EVE}{DID} = .TALKTALK\ldots$$

21.
```
H A P P Y
H A P P Y
H A P P Y
  D A Y S
---------
A H E A D
```

22. $\sqrt{CAREER} = RUT$

23. A little game of poker was played the other night, and the card filling the winning hand is represented by the letter A.

$$(A)(SPADE) = FLUSH$$

24. The card game above prompted one reader to submit the following hand in an unheard-of card game.

$$(A)(FLUSH) = TRUMPS$$

25. A plaintive cry for a long-gone TV program.

```
         C A R
 5 4 ) W H E R E
       * * *
       -----
         * * *
         * R U
         -----
           * * *
           * * *
           -----
```

26.
```
  F O O D
    F A D
---------
D I E T S
```

27. This one is to be taken literally. If you are new at alphametics, solve this one later.

```
      T H I S
          I S
      -------
    * * T O O
  H A R D *
  -----------
  * * * * * *
```

28. This may be one of the largest alphametics that still retains the distinction of having but one solution in the base-10 system.

```
                S U N
              L O S E
            U N T I E
          B O T T L E
        E L I S I O N
      N I N E T E E N
    N O N E N T I T Y
    E B U L L I E N T
    I N S O L U B L E
  -------------------
  N E B U L O S I T Y
```

29. Completely hidden cryptarithms are quite a rarity. This one has a unique solution and is really not too difficult. Note the decimal point.

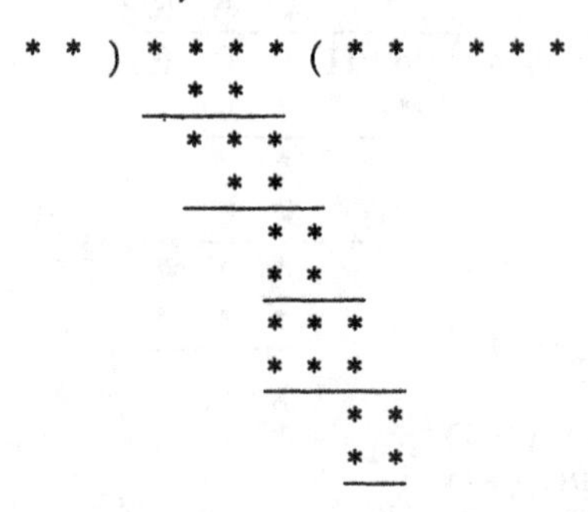

30. A good deal of space was devoted to the TIRED variation of this particular alphametic, but here is the original.

```
   T H E S E
   T E A S E
   T R I E D
 -----------
 R E A D E R
```

The next three alphametics should give some practice in working with numbers in systems other than base 10.

31. Base 11.

```
   W H E A T
   F I E L D
 -----------
 F A R M E R
```

32. Readers of *Mad* magazine will be familiar with the following character. This puzzle is to be solved in the base-9 system.

A L F R E D ÷ E = N E U M A N

33. Strangely enough, there is a solution in the base-9 system for the same name set up as a multiplication problem.

(A L F R E D)(E) = N E U M A N

34. Shakespeare did not quite say these exact words, but then he was not a recreational mathematician.

```
   A L L S
   W E L L
   T H A T
   E N D S
 ---------
 S W E L L
```

35. Hunter created this little alphametic, stating that it was the shortest true one extant. Presumably, Di (short for Diane) was a copy writer.

```
   A D
   D I
 -----
 D I D
```

Is it the shortest possible alphametic? I submit the following tale: Once upon a time a San Francisco policeman found a lost Chinese lad. "Well, my fine boy, what might your name be?" To which the frightened Chinese lad replied:

36.
$$\begin{array}{r} \text{I} \\ \text{I S} \\ \hline \text{S O O} \end{array}$$

The policeman took him to a police station. There, his sergeant located the parents, who rushed to the station, where the lad waited.

The policeman asked him, "Now, aren't you glad you've been found?" To which the happy little Chinese boy replied:

37.
$$\begin{array}{r} \text{O} \\ \text{I} \\ \hline \text{I S} \end{array}$$

ANSWERS

1.
$$\begin{array}{l} 18\ )\ 5{,}130\ (\ 285 \\ \quad\ \ \underline{3\ 6} \\ \quad\ \ 1\ 53 \\ \quad\ \ \underline{1\ 44} \\ \qquad\quad 90 \\ \qquad\quad \underline{90} \end{array}$$

2.
$$\begin{array}{r} 7{,}483 \\ 7{,}455 \\ \hline 14{,}938 \end{array}$$

3.
$$\begin{array}{r} 95 \\ 961 \\ 9{,}664 \\ \hline 10{,}720 \end{array} \qquad \begin{array}{r} 93 \\ 971 \\ 9{,}776 \\ \hline 10{,}840 \end{array}$$

4.
```
  103,295
 −95,428
   7,867
```

5.
```
5,157
1,577
   38
2,804
9,576
```

It was stipulated that the solution yielding the largest amount of CASH be found. If this restriction is lifted there are 5 additional solutions:

```
1,215   2,124   1,419   4,243   2,124
2,155   1,244   4,199   2,433   1,244
   86      68      75      57      68
3,694   3,809   2,503   1,706   5,807
7,150   7,245   8,196   8,439   9,243
```

6. 7,852 ÷ 4 = 1,963. If the restriction, the odd GIFT, is lifted, another solution is 6,952 ÷ 4 = 1,738.

7. *a.*
```
  148
  874
1,022
```
b.
```
 2,482
 8,364
10,846
```

8.
```
 21,469
 85,463
106,932
```

9.
```
 9,552
   902
   382
10,836
```

10.
```
   312      214
   921      941
    73       52
 9,273    9,452
10,579   10,659
```

11.
427
27
2,989
854
11,529

12.
621
846
9,071
10,538

13.
106
19,722
82,524
102,352

14.
930
58,682
878,682
938,294

It was hinted that TWO and TWENTY are even numbers. However, there is another solution if this restriction is ignored: TWO + SEVEN + ELEVEN = TWENTY becomes

$$923 + 58{,}784 + 868{,}784 = 928{,}491$$

15.

35	34
35	34
8,230	8,632
84,372	87,316
92,672	96,016

16.
6,964
13,244
84,546
104,754

Harry L. Nelson, of the Lawrence Radiation Laboratory of the University of California, pointed out that in base 17 the addition 9 + 3 + 7 = 12 is true. If this alphametic is solved in base-17 arithmetic, however, there are 340 distinct solutions. For example:

(14) I (14) 3 + 1 2 R 3 3 + (15) 3 4 3 (14) = 1 0 3 L 4 3

with 21 choices for R, I, and L and with R less than I. This alone gives 42 solutions, while similar results for other values will give the other 298 solutions.

17.
```
 824,163     631,708
 +91,573     -57,428
 -------     -------
 915,736     574,280
```

18.
```
    3,462
      546
---------
   20 772
  138 48
 1 731 0
---------
1,890,252
```

19.
```
388 ) 91,180 ( 235
      77 6
      -----
      13 58
      11 64
      -----
       1 940
       1 940
       -----
```

20. $\frac{242}{303} = .79867986\ldots \qquad \frac{212}{606} = .34983498\ldots$

21.
```
29,661
29,661
29,661
 3,910
------
92,893
```

22. $\sqrt{376{,}996} = 614$

23. $(5)(13{,}582) = 67{,}910$

Since the statement given says that the winning hand (FLUSH) was filled by a card represented by A, the total of 6 letters in "A FLUSH" must represent only 5 cards. Since A cannot equal 1, one of the letters in FLUSH must be equal to 1. Then the solution given means that the 5, 6, 7, 9, and 10 of spades was the flush hand. If the use of 12 is permitted to signify a Queen, then another solution is possible: $(4)(17{,}453) = 69{,}812$.

There are a number of other solutions when the clues are not considered to be restrictions. Derrick Murdoch, a Canadian alphametics enthusiast, noted that A can be 2 (with SPADE = 38,215; 39,215; 35,218; or 35,219, and FLUSH = 76,430; 78,430; 70,436; 70,438) or 4 (with SPADE = 17,453, and FLUSH = 69,812).

24. $(6)(45{,}183) = 271{,}098$

25.

$$\begin{array}{r} 257 \\ 54\overline{)13{,}878} \\ \underline{10\,8} \\ 3\,07 \\ \underline{2\,70} \\ 378 \\ \underline{378} \end{array}$$

26.

$$\begin{array}{r} 9{,}551 \\ \underline{931} \\ 10{,}482 \end{array}$$

27.

$$\begin{array}{r} 4{,}379 \\ \underline{79} \\ 39\,411 \\ \underline{306\,53} \\ 345{,}941 \end{array}$$

28.
```
          741
        5,672
       41,982
      369,952
    2,587,861
   18,129,221
  161,219,890
  234,558,219
  817,654,352
-------------
1,234,567,890
```

29.
```
16 ) 1,062 ( 66.375
       96
      ---
      102
       96
      ---
        60
        48
       ---
       120
       112
       ---
         80
         80
         --
```

30.
```
 53,646
 56,046
 51,269
-------
160,961
```

31.
```
 95,307     97,305
 1X,328     18,32X
-------    -------
104,634    104,634
```

H and I, and T and D are interchangeable; $X = 10$, a single digit in the base-11 system of numeration.

32. $704{,}836 \div 3 = 231{,}572$ (Base 9)

33. (164,057)(5) = 852,318 (Base 9)

34.
$$\begin{array}{r} 9{,}332 \\ 8{,}433 \\ 6{,}596 \\ \underline{4{,}072} \\ 28{,}433 \end{array}$$

35.
$$\begin{array}{r} 91 \\ \underline{10} \\ 101 \end{array}$$

36.
$$\begin{array}{r} 9 \\ \underline{91} \\ 100 \end{array}$$

37.
$$\begin{array}{r} 9 \\ \underline{1} \\ 10 \end{array}$$

8

conglomerate

A number of topics in recreational mathematics, though interesting in their own light and with a rich history and tradition, are not extensive enough for a chapter to be devoted to them. Several of these minor topics are dealt with here, and the great range of ideas that has been traditionally studied in the field of recreational mathematics is indicated.

THE PROBLEM OF THE MICE

Four mice were at the four corners of a square room. For want of a cat to chase them, they would chase each other. One day, a mouse decided to chase the neighbor on his right, and it happened that this same thought occurred simultaneously to each mouse. The mice all moved at the same constant speed and always headed directly toward their moving objective. Which mice met first? Where did they meet? How far did they travel? The situation at about the middle of the chase is shown in Figure 113*a* and the completed chase is shown in 113*b*.

The final paths taken by the mice are spirals and the length of each spiral is the same as the length of a side of the square room. There are a number of proofs confirming the path and the length, but here is a fairly simple geometry proof, involving some elementary trigonometry, which was supplied by Professor Leo Moser of the Department of Mathematics at the University of Alberta in Edmonton, Alberta.

Consider the case of a square with sides equal to unity. A and B are 2 adjacent mice; A chases B, and both move at a constant velocity of 1.

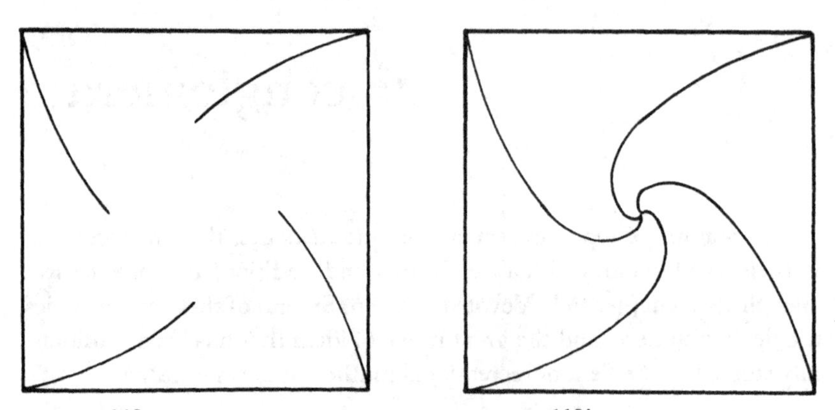

FIGURE *113a*

FIGURE *113b*

Now A is going toward B with velocity 1. On the other hand, it can be seen that the component of velocity of B in the direction away from A is the cosine of 90°, or 0. Hence, the side of the square (at any given time the mice are at the corners of a square) is decreasing with a velocity of (1 − cosine 90°), and, since it starts out with sides equal to 1, it will become a square with sides equal to 0 after a time $\frac{1}{(1 - \text{cosine } 90^\circ)}$. This last expression is equal to 1, and it is also the total length of the path of one mouse.

It is interesting to consider other regular polygons with mice in the corners. If 2 mice are at the ends of a narrow hall (a "2-sided" polygon) it is no great mental effort to deduce that if they decide to chase each other they will collide head-on just halfway down the hall, assuming simultaneous starts and equal and constant velocity. If 3 mice are at the vertexes of an equilateral triangle they will describe spirals, but travel only ⅔ the length of a side of the triangle before the first collision.

What about a pentagonal room, or a higher-order polygonal room? Professor Moser provides a simple method of calculating the distance the mice travel, assuming the lengths of the sides of the polygon are unity, the starts are simultaneous, and the velocities are constant and equal. If A and B are 2 adjacent mice in an n-sided polygonal room with A chasing B, the component of velocity of B in the direction away from A is the cosine of $\frac{360°}{n}$. The side of the polygon is decreasing with a velocity of $\left(1 - \text{cosine}\,\frac{360°}{n}\right)$. The reciprocal of this last expression is the path length of a mouse. The following table shows the path lengths for mice chasing each other in regular polygonal rooms, for path length, l, in a regular n-sided polygonal room:

n	l	n	l	n	l
1	∞	6	2.	20	20.4332
2	0.5	7	2.6560	50	126.744
3	0.6667	8	3.4143	100	507.614
4	1.	9	4.2742	1000	50,660.87
5	1.4472	10	5.2363	∞	∞

The formula yields the same answer for a one-sided room and an infinite-sided room. Actually, the physical interpretation is not the least bit difficult, if one allows his imagination a little freedom. In a "one-sided" polygonal room, there would be but one mouse, and this poor fellow would have only himself to chase. That could take a long time. As for the infinite-sided room, imagine an infinite number of mice stationed along the wall of a large roundhouse, an "infinite-sided" polygonal room. At a signal, each mouse turns and pursues his right-hand neighbor. Obviously, these creatures are doomed to run around one large circle forever, never approaching the center of the room, never catching each other.

The problem can be reversed, leading to some startling results. For example, in what kind of regular polygonal room would mice have to be so that each travels exactly 3 units before catching his neighbor? The formula yields the very logical answer: a room with approximately 7.47 sides, of course. The physical interpretation of 7.47 mice chasing each other in a 7.47-sided room I leave to the talented reader.

The analysis of irregular polygonal rooms of more than 3 sides is too complicated to discuss here, but the velocity relationships for a *scalene triangle,* one in which the three sides are unequal, have been known for some time. If 3 mice start from the vertexes of such a triangle so that A chases B, B chases C, and C chases A, with initial distances between pursuer and pursued of c, a, and b, respectively, they will meet simultaneously if their relative velocities are proportional to b^2c, c^2a, and a^2b, respectively. For certain triangles, one mouse may suffer the rather upsetting experience of heading almost directly toward his pursuer when almost ready to capture his own quarry.

DEPLOYMENT

The reader may wish to know of a game similar to, but not quite as elementary as, ticktacktoe. A college student, William H. McGrail, of Worcester, Massachusetts, some years ago also felt the need for a new, silent, truly competitive pencil-and-paper game. The result of Mr. McGrail's inner need is the following amusement, which is called "deployment."

The game, in the simplest form offering sufficient challenge, is played on a 5 × 5 array of 25 squares by two players. As will become apparent during the explanation, a 3 × 3 or 4 × 4 array produces rather short games that hardly surpass ticktacktoe in complexity. In any one of the squares, the first player puts a figure "1"; the second player places, in another square, the figure "2"; the first player places a figure "3" in another square; the second player places a figure "4" in another square.

The first player starts again with the figure "1" and the game continues as before until each player has put down 2 groups of 6 numbers each—filling all but one of the array of squares. Points are awarded on the basis of the final arrangement of the figures on the squares and the high score wins. The following scale is used for scoring:

5 figures in a row	6 points
4 figures in a row	5 points
3 figures in a row	4 points
6 figures in a chain	3 points
5 figures in a chain	2 points
4 figures in a chain	1 point

A *row* is defined as any configuration of 3 or more like squares (those containing the same figure or number) lying in a straight line horizontally, vertically, or diagonally. A *chain* is defined as any configuration of 4 or more like squares linked to each other by at least one corner or edge. Rows and chains are counted separately, even though both may have common squares.

An illustrative game is shown in Figure 114:

	f	g	h	i	j
a	4	1	4		2
b	1	4	1	1	1
c	3	4	2	2	3
d	4	4	3	3	2
e	3	3	2	2	1

FIGURE *114*

1. 1–*bh*	13. 1–*ag*
2. 2–*ch*	14. 2–*dj*
3. 3–*di*	15. 3–*eg*
4. 4–*bg*	16. 4–*ah*
5. 1–*bi*	17. 1–*bf*
6. 2–*ci*	18. 2–*ei*
7. 3–*cj*	19. 3–*cf*
8. 4–*cg*	20. 4–*df*
9. 1–*bj*	21. 1–*ej*
10. 2–*eh*	22. 2–*aj*
11. 3–*dh*	23. 3–*ef*
12. 4–*dg*	24. 4–*af*

The scoring for this game is as follows:

Odd Team			*Even Team*		
1	3 in a row	4	2	less than 3 in a row	0
	5 in a chain	2		5 in a chain	2
3	less than 3 in a row	0	4	3 in a row	4
	5 in a chain	2		6 in a chain	3
	Total	8		Total	9

Each of the 4 groups is usually awarded points for one or more rows, and for a chain. Rows within a group that cross in different directions and utilize a common square are counted as separate ones (see Figure 115*c*). Thus, we may have a chain with no rows in it, as in Figure 115*a*; a row or rows with no chains, as in 115*b*; a row or rows and a chain together, as in 115*c*; or a configuration with no rows or chains at all, as in 115*d*. The last is an example of a hypothetical scoreless position that might result when a game is played.

This game may be expanded indefinitely to allow for any number of players or grouping of teams. The $n \times n$ array of squares must be of the form $2N + 1 = n^2$, where $2N$ is the total number of figures. Also, $\frac{N}{n+1}$ is the number of teams of two figures each, and $n + 1$ is the num-

FIGURE *115a*

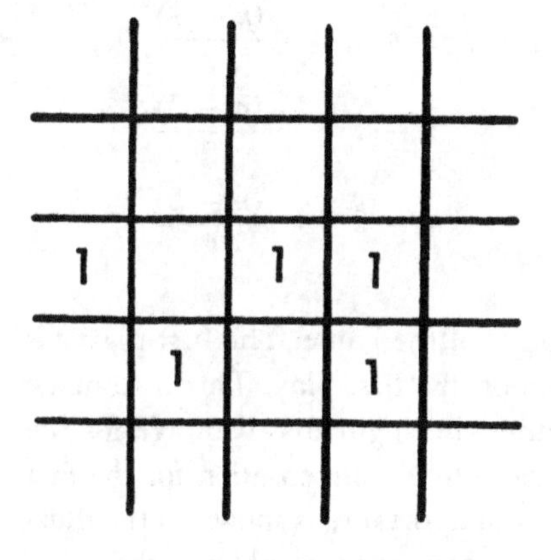

FIGURE *115b*

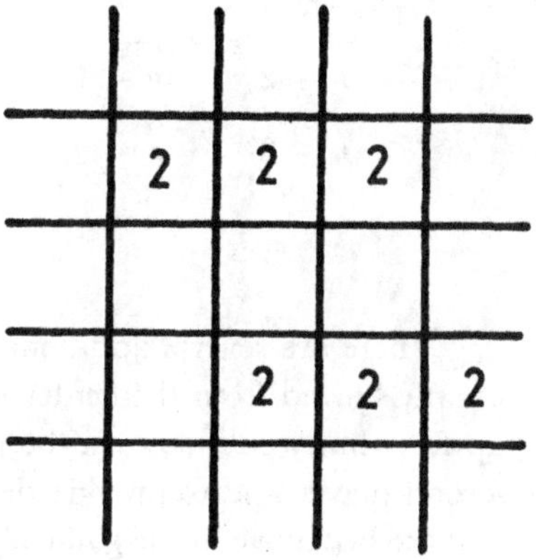

FIGURE *115c*

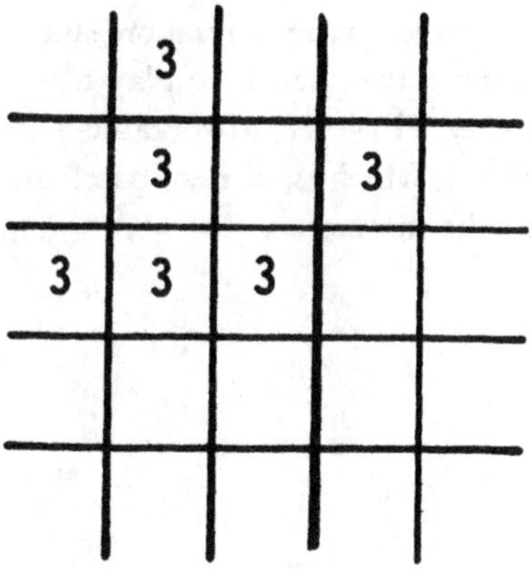

FIGURE *115d*

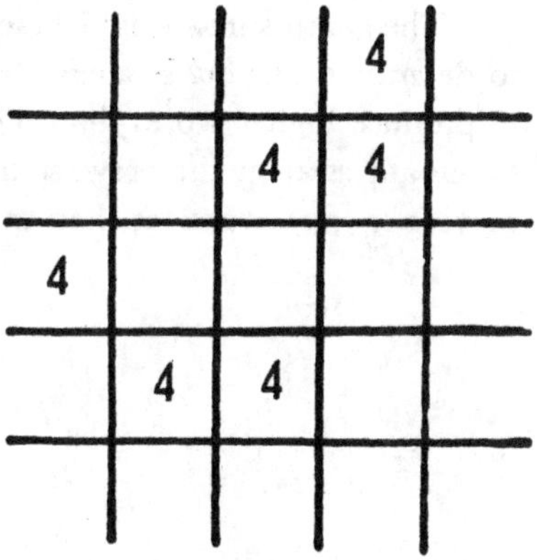

ber of figures in a group. Point values operate according to a scale as well:

In a Row	*Points*	*In a Chain*	*Points*
n	$n + 1$	$n + 1$	$\frac{(n + 1)}{2}$
$n - 1$	n	n	$\frac{(n - 1)}{2}$
$n - 2$	$n - 1$	$n - 1$	$\frac{(n - 3)}{2}$
$n - 3$	$n - 2$	$n - 2$	$\frac{(n - 5)}{2}$

There are several additional aspects of the game. The first player is usually barred from the center square on the first play. This is to make up for what would seem to be an otherwise slight disadvantage for the second player and to provide the latter with a compensation for the fact that the last move in the game is often unimportant, as shown in the illustrative game. In case of a game between three or more players, the privilege of playing the center square could be accorded the last player only during the first round of play, although this may not be mathematically justified.

The game, surpassing ticktacktoe in complexity, requires more study to determine playing strategy. It does appear that defensive play tends to produce higher scores than offensive play. However, a workable and systematic strategy may reverse this tendency, which is, at best, based on playing more at random than in a methodic manner.

DOMINO RECREATIONS

The game of dominoes will not be discussed; rather, some recreations involving the use of dominoes and the patterns they may take will be developed.

If dominoes are treated as markers, and the number of pips, or dots, are used as numerical equivalents, some simple arithmetical problems may be set up using them. The number of dots on each half of a domino is considered as a single number. A few such arithmetic problems are shown in Figure 116. These are equivalent to the following multiplication and addition problems.

500	415	554
2	4	6
1000	1660	3324

	12	
	2	
	42	33
6	55	536
6	515	653
12	0626	1222

A variation is to use a complete set of dominoes for a problem set-up For example, a double-4 set of dominoes contains 15 pieces, and these can be used in 3 groups of 5 each to form sums of 21, 22, and 23 (see Figure 117). Figure 118 shows all 28 dominoes of the double-6 used in a single addition problem.

Another major domino recreation is the formation of like numbers of dots into groups of four. The French mathematician Edouard Lucas

FIGURE *116*

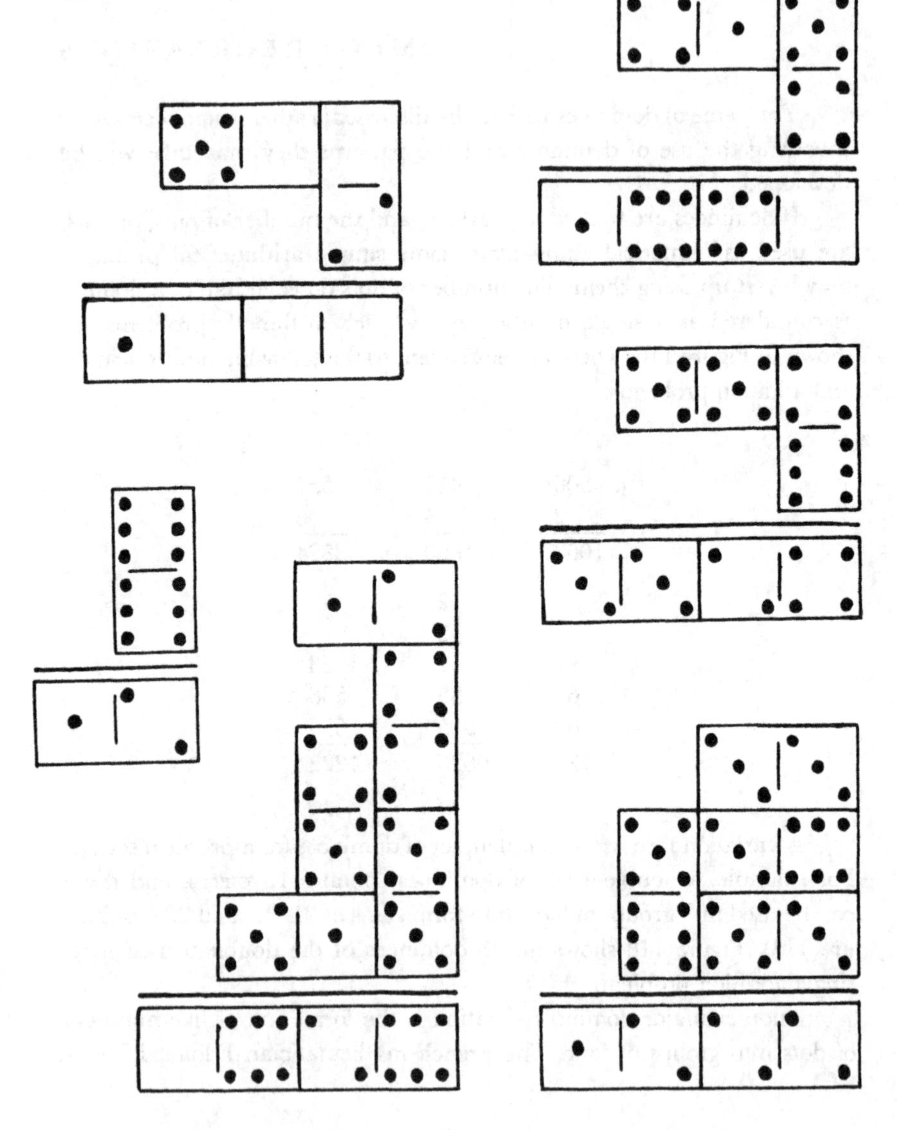

(1842–1891) gave the name *quadrilles* to such formations. By considering the properties of dominoes, it is possible to determine the sets from which quadrilles may be assembled. If the largest number of dots on a domino

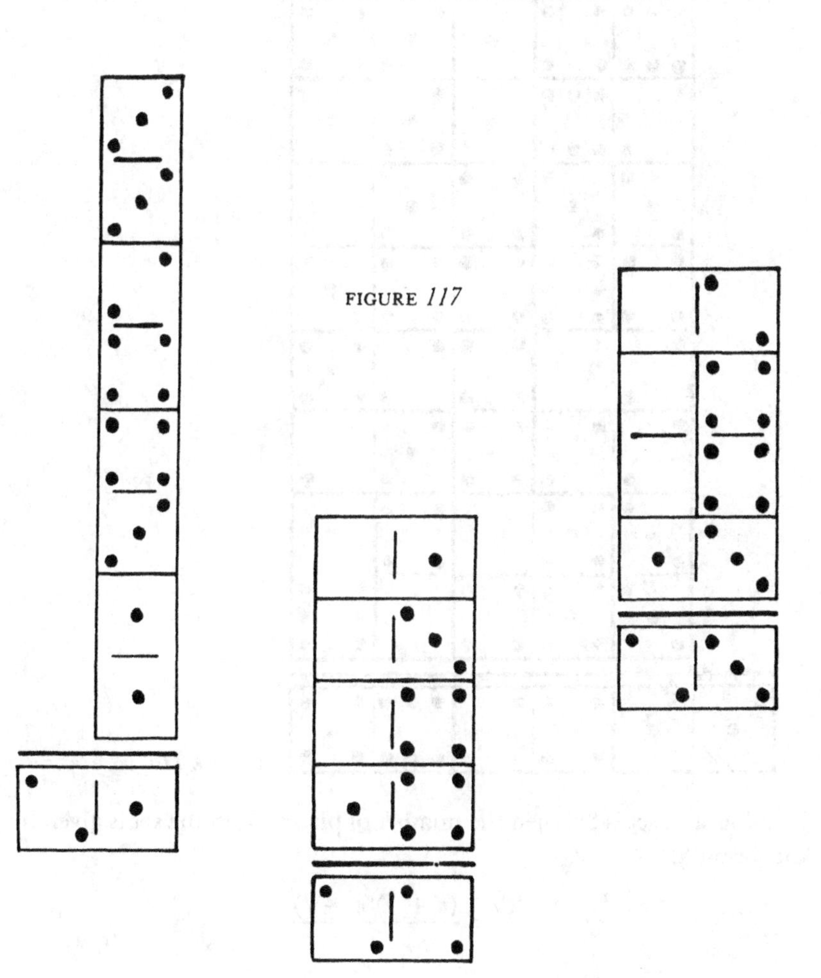

FIGURE 117

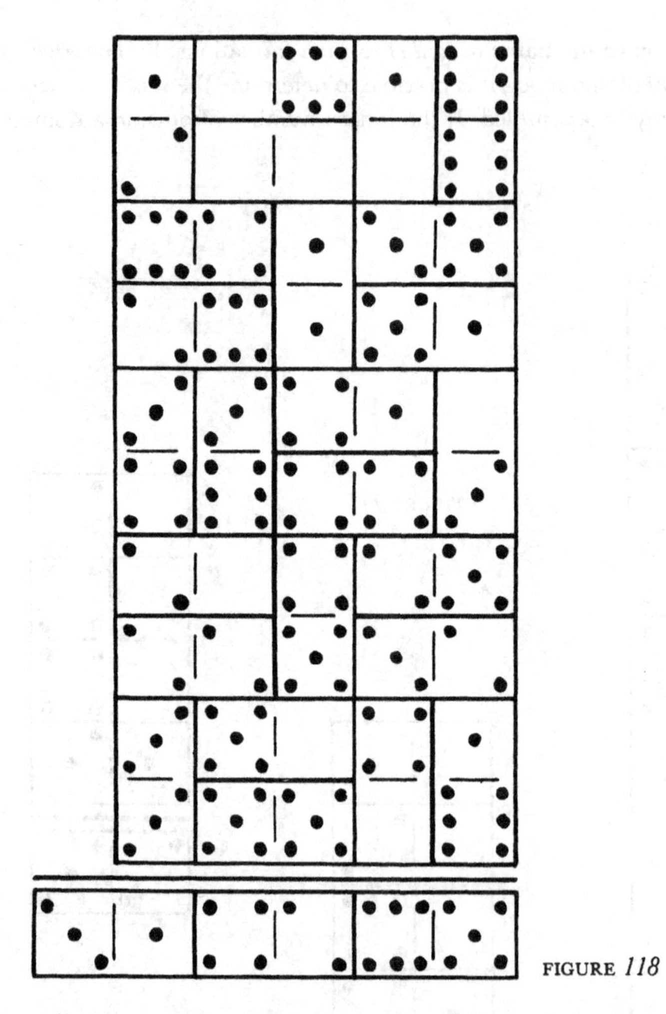

FIGURE *118*

in a double-n set is $2n$, then the number of pieces, N, in the set is given by the formula:

$$N = \frac{(n + 1)(n + 2)}{2}$$

The total number of dots in a complete set of dominoes is given by $D = nN$, where D is the number of dots in the set, and the table lists the result:

n	N	D	n	N	D	n	N	D
0	1	0	7	36	252	14	120	1,680
1	3	3	8	45	360	15	136	2,040
2	6	12	9	55	495	16	153	2,448
3	10	30	10	66	660	17	171	2,907
4	15	60	11	78	858	18	190	3,420
5	21	105	12	91	1,092	19	210	3,990
6	28	168	13	105	1,365	20	.231	4,620

Each number, from zero to n, appears in a set $n + 2$ times. A necessary condition for the formation of quadrilles is that each number appear in the set a multiple of four times. This may be stated as

$$\frac{n + 2}{4} = k$$

where k is an integer. This equation can be solved for n,

$$n = 4k - 2$$

where k is any positive integer. The possible sets of dominoes that can be used for the formation of quadrilles are those given by the series 2, 6, 10, 14, 18 . . . .

It is not difficult to find a solution for the 6 dominoes in the double-2 set (see Figure 119). It can be shown that there is only one distinct answer in this case. Each distinct solution is a particular pattern of dominoes and gives rise to others by the permutation, or rearrangement, of the numbers for a total of $(n + 1)!$ answers. A solution is considered distinct if *some* of the cells (blocks of four numbers) containing a given pair of numbers

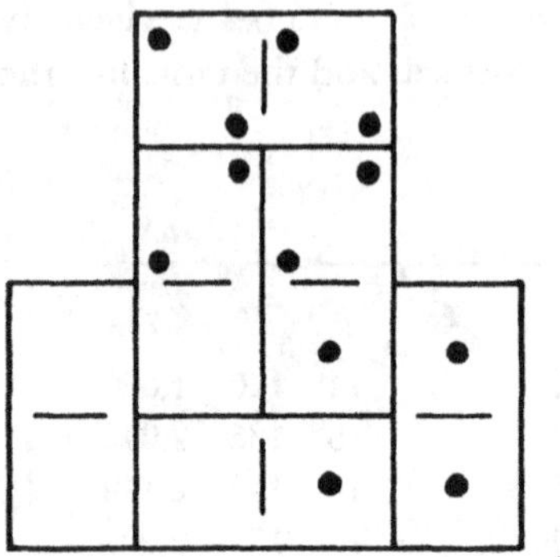

FIGURE *119*

FIGURE *120a*

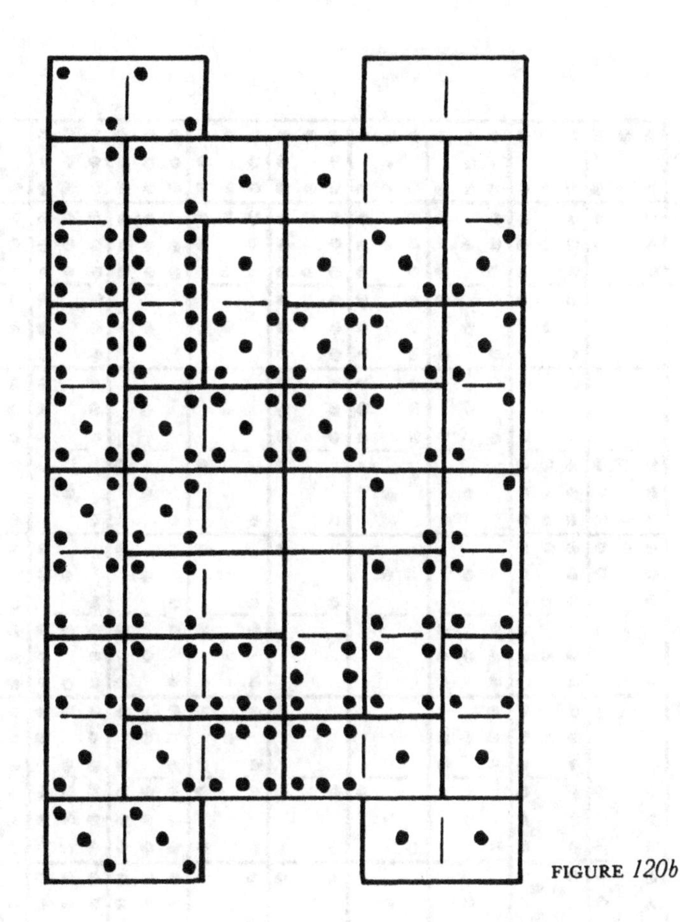

FIGURE *120b*

are interchanged. If *all* cells involving a pair of numbers are interchanged the rearrangement is but one of the $(n + 1)!$ permutations and is not considered distinct.

Seven quadrille solutions were published in 1891 in Volume II of Edouard Lucas' *Récréations Mathématiques,* a four-volume collection of

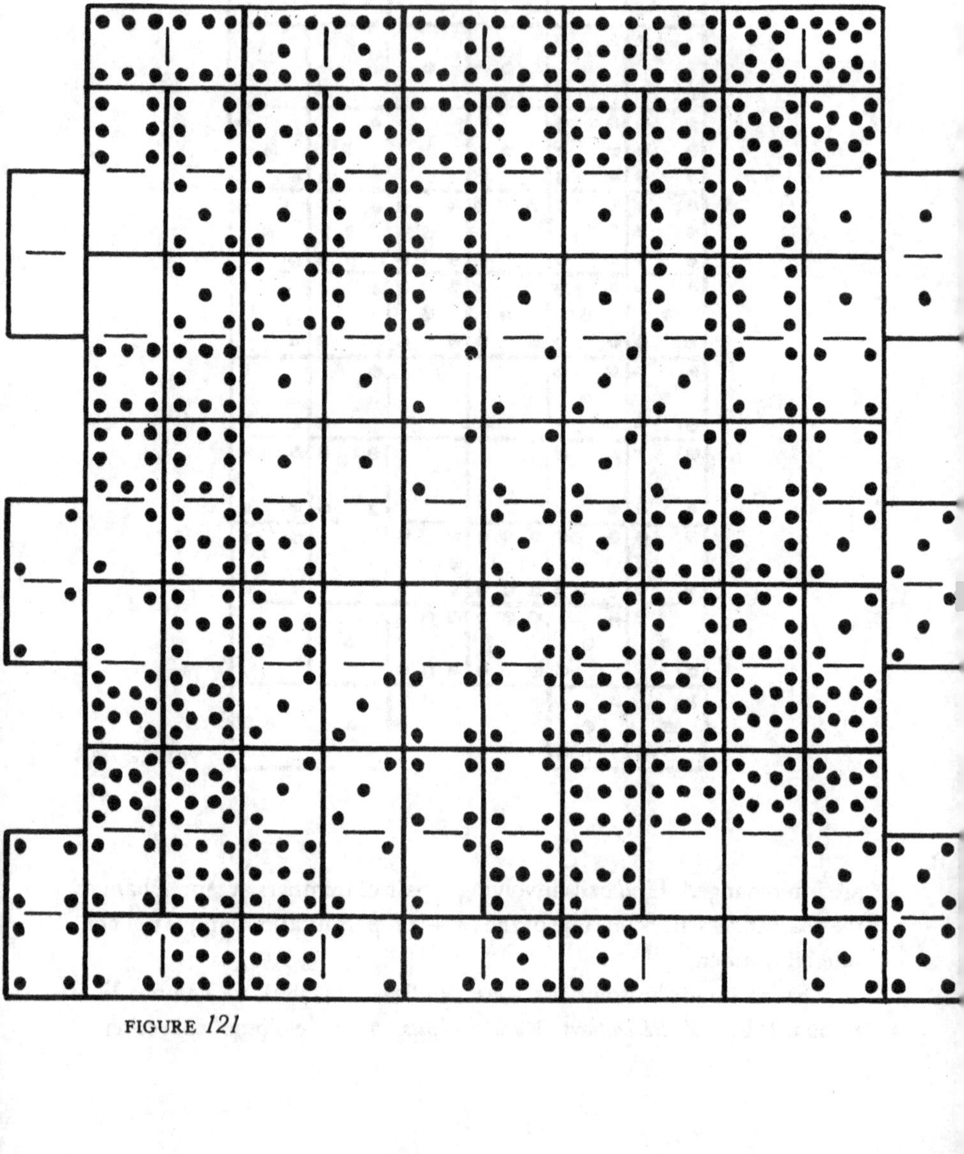

FIGURE *121*

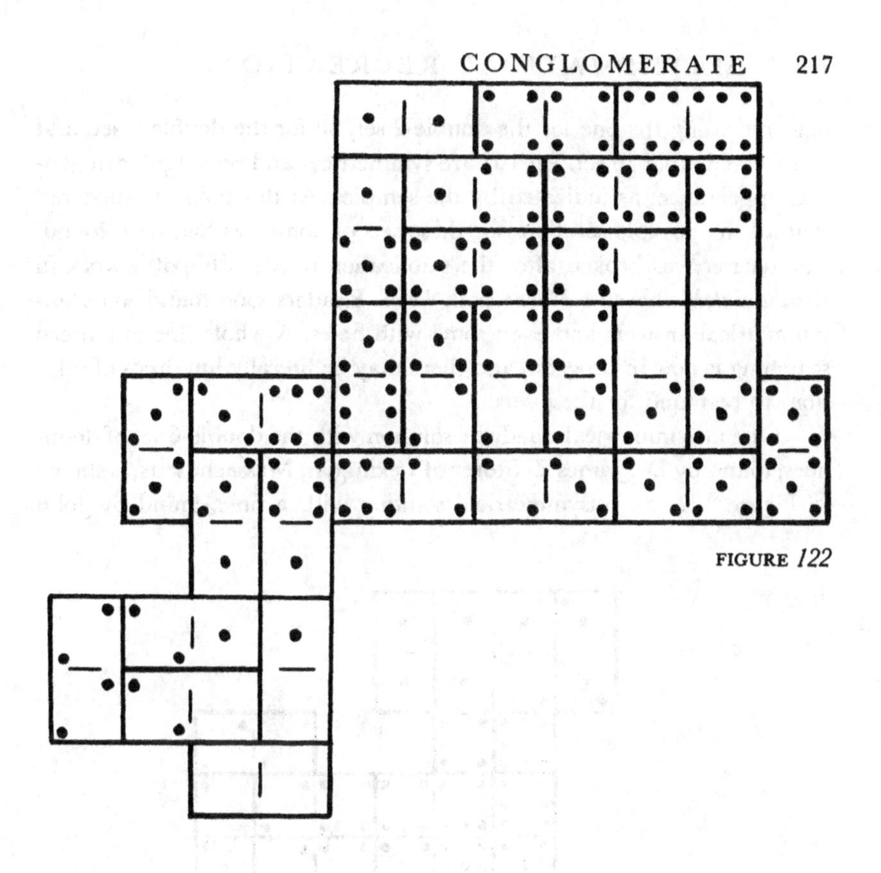

FIGURE *122*

mathematical recreations. All the solutions were for the 28 dominoes of the double-6 set. Recently, Wade E. Philpott, an inveterate domino puzzlist of Lima, Ohio, found additional solutions for the double-6 set of dominoes and a number of solutions using the 66 dominoes in the double-10 set. Two of the solutions given by Lucas are shown in Figure 120, and a solution by Philpott for the double-10 set of dominoes is shown in Figure 121.

Philpott diligently managed to catalog 104 distinct quadrille solutions including the one for the double-2 set, 58 for the double-6 set, and 45 for the double-10 set. All 104 are symmetrical and rectangular in general appearance, as indicated by the samples. At this point, it appeared that all the possible solutions for these sets of dominoes had been found. The barrier was broken after the publication of Mr. Philpott's work in *Recreational Mathematics Magazine* in 1964. Readers soon found some unsymmetrical answers and even some with holes. A whole line of pattern searching is now in progress, and there may be literally hundreds of solutions to be found for these sets.

One unsymmetrical quadrille solution with the double-6 set of dominoes, found by Dr. James E. Storer of Lexington, Massachusetts, is shown in Figure 122; an unsymmetrical solution with a hole, found by John

FIGURE *123*

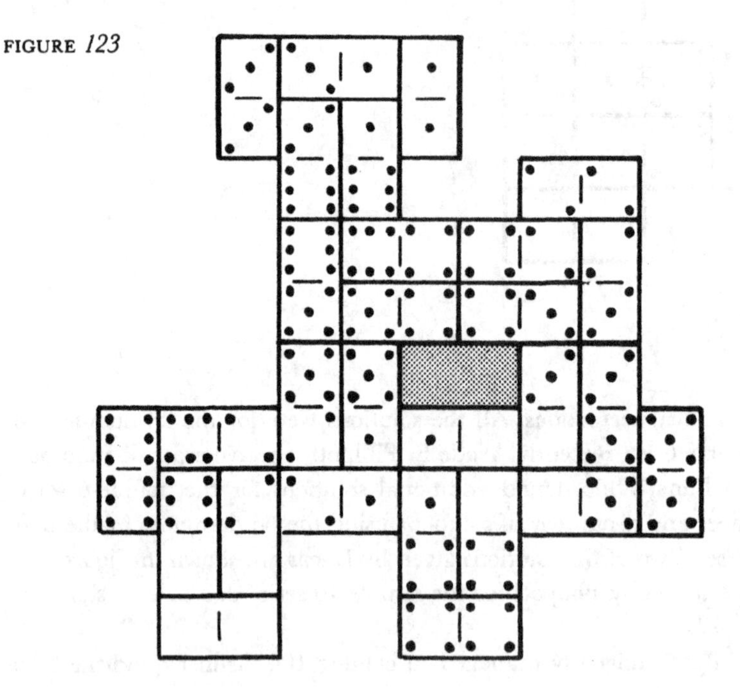

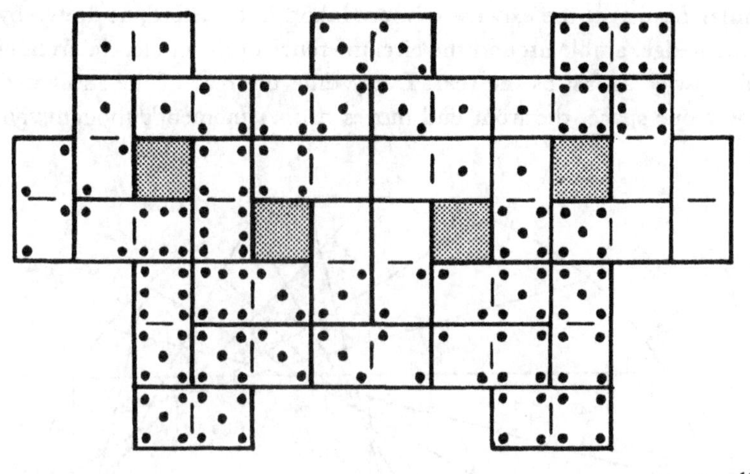

FIGURE *124*

Haliburton of Guymon, Oklahoma, is shown in Figure 123; and a symmetrical 4-hole solution by Storer is shown in Figure 124.

Philpott has evolved a method of constructing a quadrille formation of any desired size, using larger and larger sets of dominoes, by a process of building up from known solutions to the lower-order sets of dominoes. That this method is successful means that there are an infinite number of quadrille patterns.

LOST CHORDS

W. H. Cozens, of Ilminster, Somerset, England, has discovered that a chord that is allowed to get "lost," that is, go wandering around its circle on the loose, can produce some fascinating results.

Draw a circle and divide its circumference into any number of equal parts. The diagrams shown here, except Figures 129 and 134, are marked out with 10° spacings, making 36 parts. Smaller divisions will produce similar results, at the expense of more labor. Let a chord, as drawn by a straightedge, amble around the circumference of the circle, the front end going twice as fast as the rear. Every time the rear of the straightedge moves one space, the front end moves two. A moment's thought shows

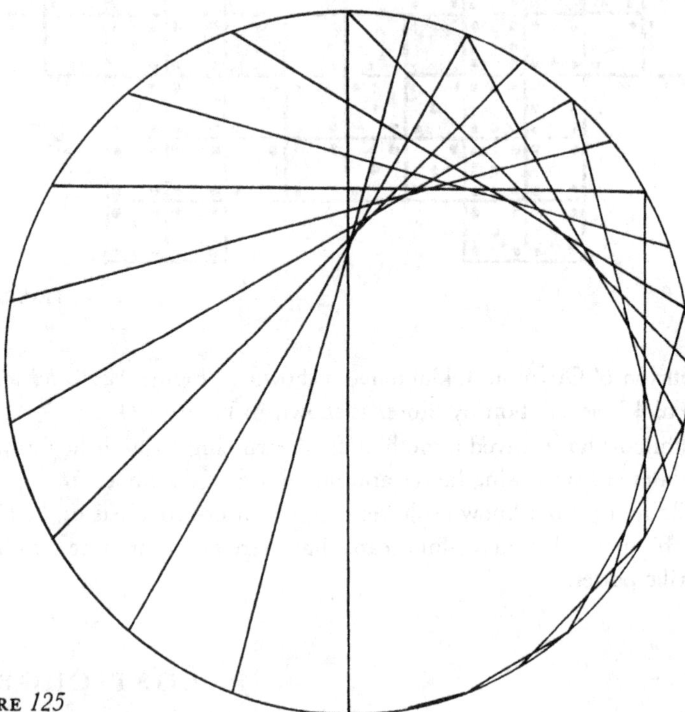

FIGURE *125*

that by the time the fast end has completed the trip around the circle, the slow end has traveled just halfway around (Figure 125). Meanwhile, although every line has been drawn with a straightedge, something strikingly like a curve seems to be emerging.

Continue exactly as before. When the slow end has traversed its full circle, the faster end will have completed its second circuit. The result is shown in Figure 126, an elegant curved cusp produced by straight lines. This particular curve has been known for centuries as the *cardioid*, or heart, curve.

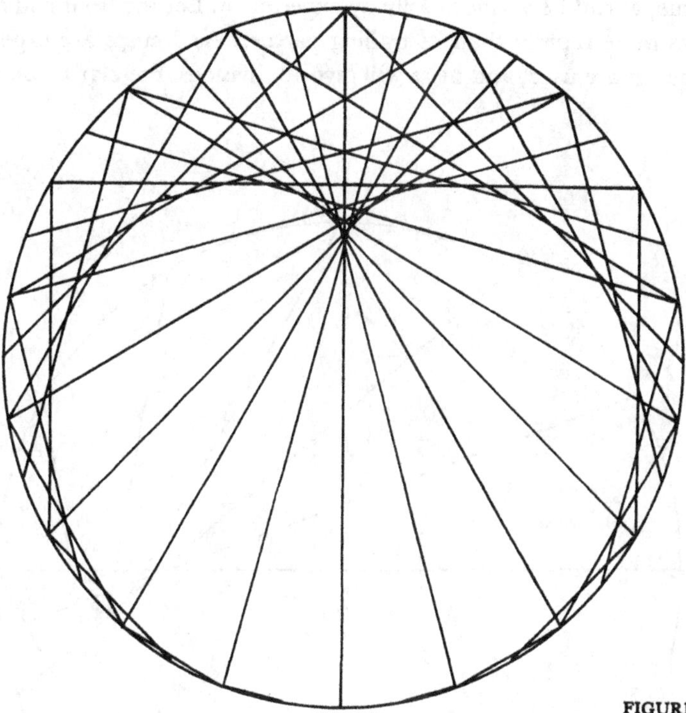

FIGURE *126*

Such striking results call for further experimentation. Draw another circle and let the leading end of the chord travel 3 times as fast as the other. When the slow end has covered 1 right angle, the fast end will have covered 3, so that the chord will have become a diameter. As the chord must now begin to shorten again, this diameter must mark the position of one cusp (Figure 127). By the time the slow end has completed its circle, the other will have done 3 laps, and a symmetrical 2-cusped curve is visible. This curve is called the *nephroid,* or kidney, curve (Figure 128).

These two results suggest that when the front end moves n times as fast as the rear, a figure with $n - 1$ cusps will be produced, and this is in fact true, as can be verified easily by experiment. Let the front end move 6 times more rapidly than its trailing partner. As 5 cusps are expected, 9° spacings are used, and these will give 40 divisions. For clarity, only the

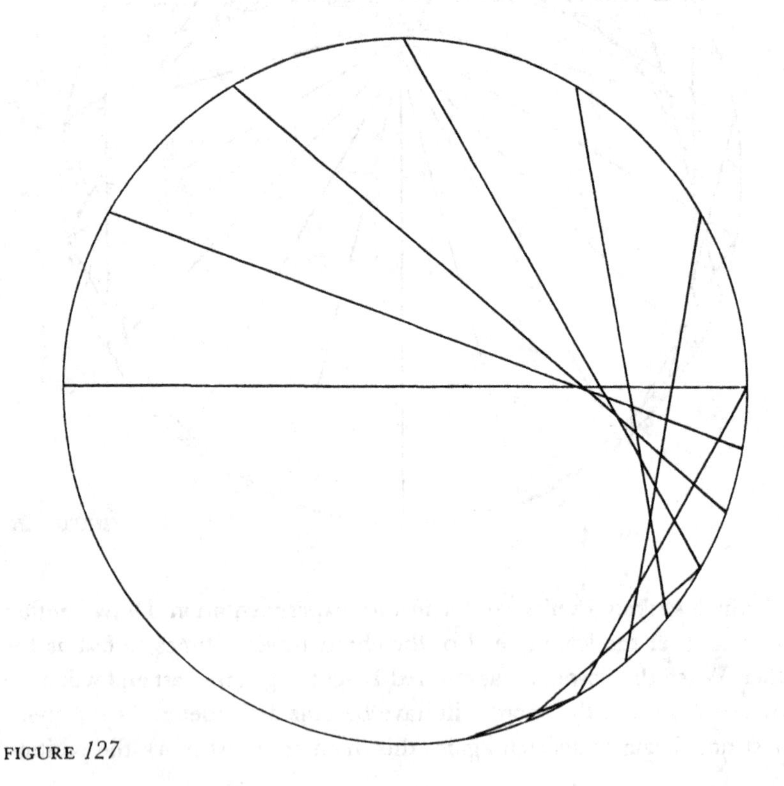

FIGURE *127*

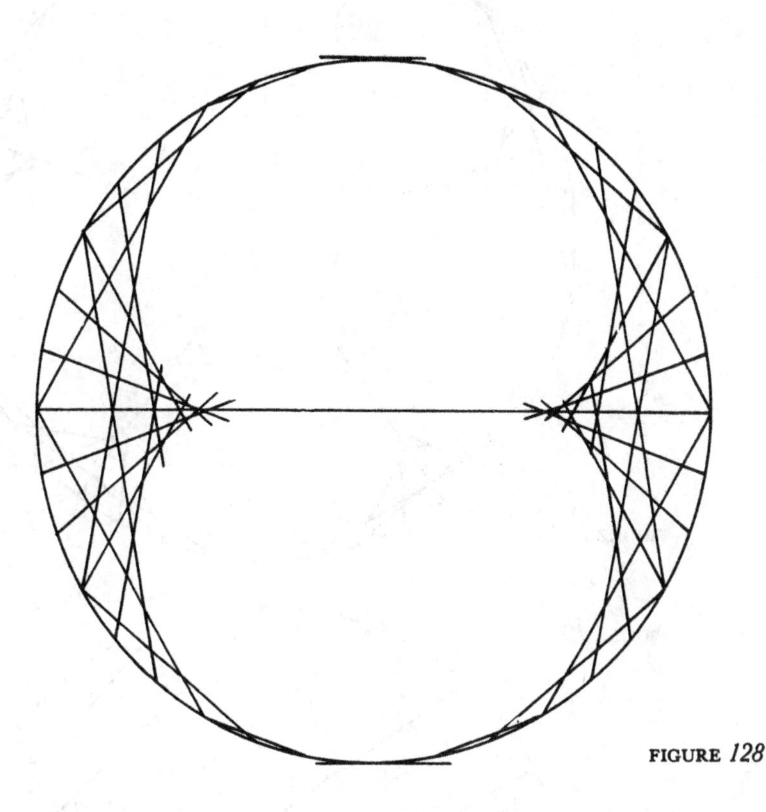

FIGURE *128*

ends of the chords have been drawn in Figure 129. The resulting 5-petaled curve has no common name, but since it looks like a buttercup it will be called a *ranunculoid,* after *Ranunculus,* the buttercup genus.

One line of experimentation leads to another. It has been established that if one end of a chord goes n times as fast around the circumference of a circle as the other end, then the chord will envelope a symmetrical curve with $n - 1$ internal cusps.

What if the second end of the chord moved in the opposite direction? It is sufficient for the first experiment along this line of research to let the

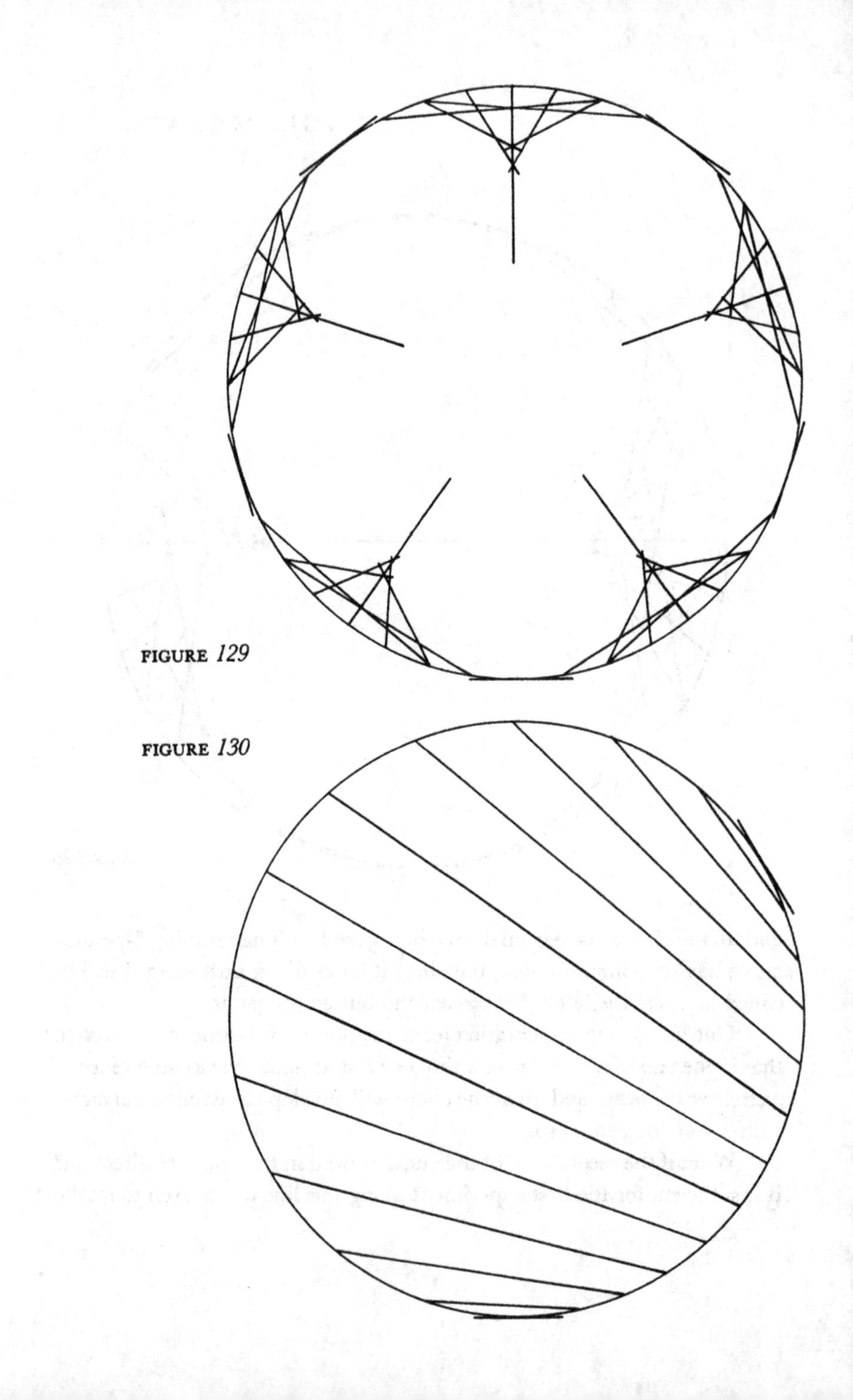

FIGURE *129*

FIGURE *130*

ends go in opposite directions, one twice as fast as the other. Figure 130 shows the somewhat uninteresting result.

If the slower end completes a revolution, the result is apparently most disappointing. There is no sign of a curve or a cusp, as Figure 131

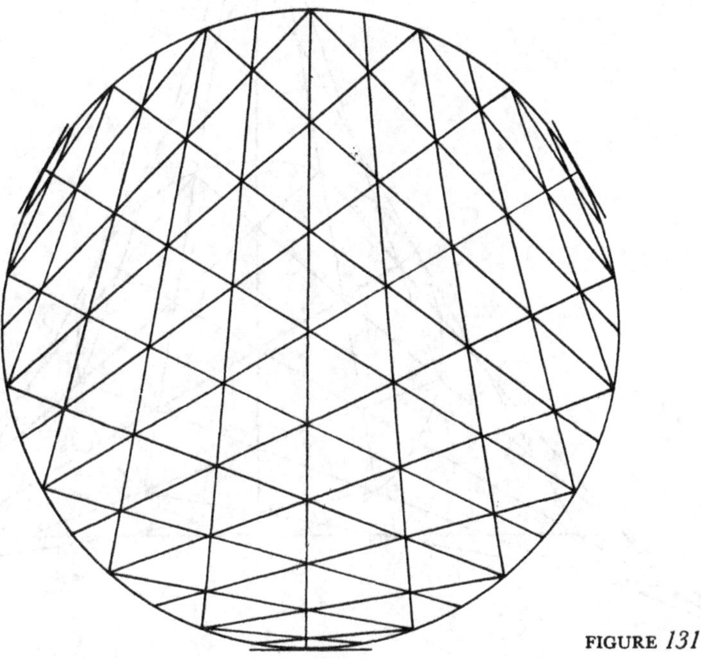

FIGURE *131*

shows, but, before dismissing this experiment as a failure, it is worth taking a closer look at the drawing. Those chords are in groups that seem to converge to the top, the bottom left, and the bottom right. What if they were extended and allowed to meet outside the circle?

The result is shown in Figure 132. Circumscribed about the circle is

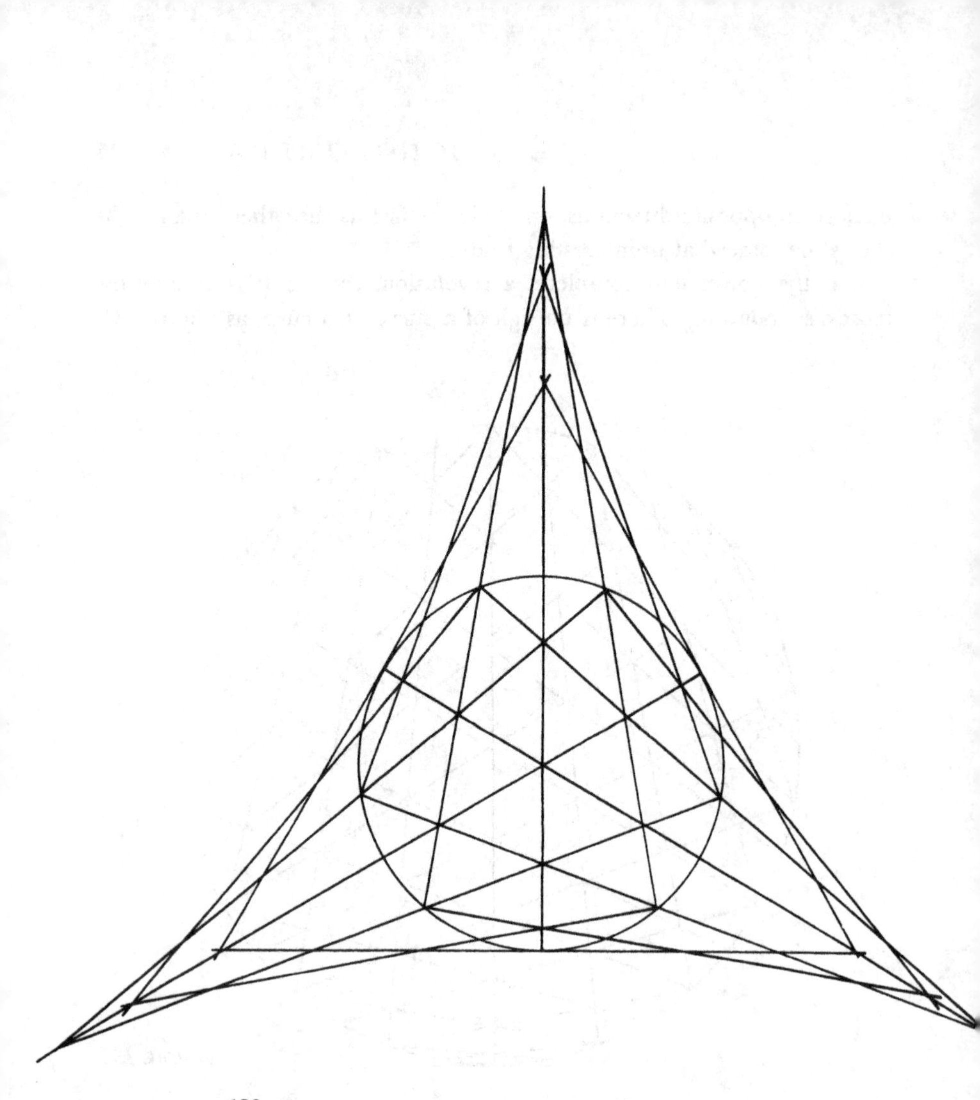

FIGURE *132*

a figure with 3 symmetrical cusps turning outward from it. This is the *deltoid,* so named, from its resemblance to the Greek letter delta, by the eighteenth-century Swiss mathematician Leonhard Euler (1707–1783).

The next experiment suggests itself and its result. Let the ends rotate in opposite directions but let one travel 3 times as fast as the other; the result ought to be a 4-cusped star. The result is indeed as expected and the *astroid,* or star, curve is shown in Figure 133.

It now appears that when one end moves n times faster than the other, but in an opposite direction, the chord envelopes, from the inside, a curve that lies entirely outside the circle. The number of cusps, however, is not $n - 1$ but $n + 1$. The assumption can be tested by letting one end move 4 times as fast as its partner. For this purpose, since 5 external cusps

FIGURE *133*

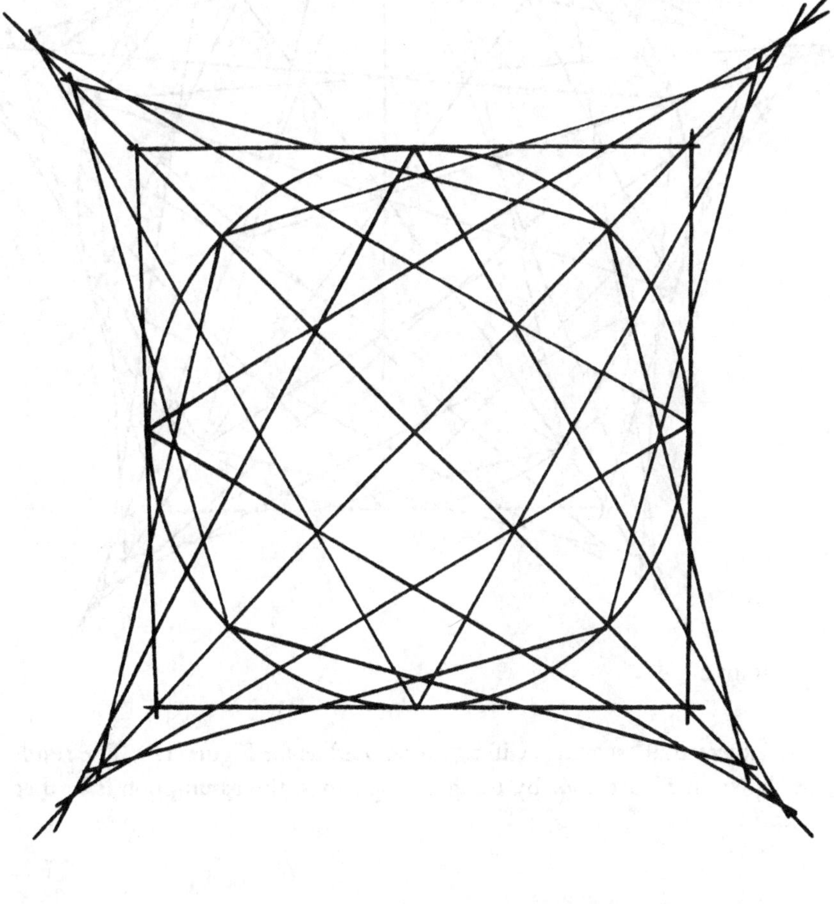

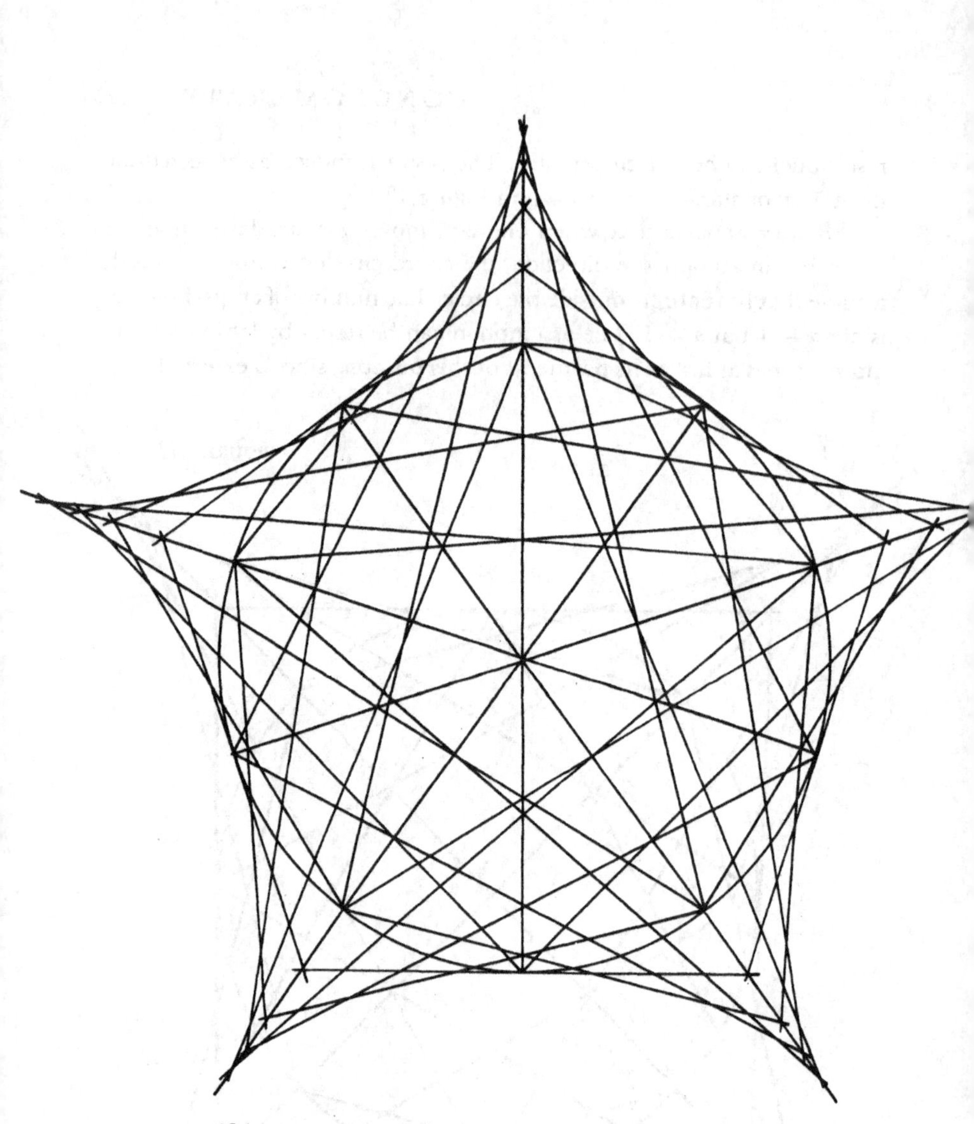

FIGURE *134*

are expected, 9° spacings will again be used as for Figure 129. The result is shown in Figure 134. By using 10° spacings, the assumption is further

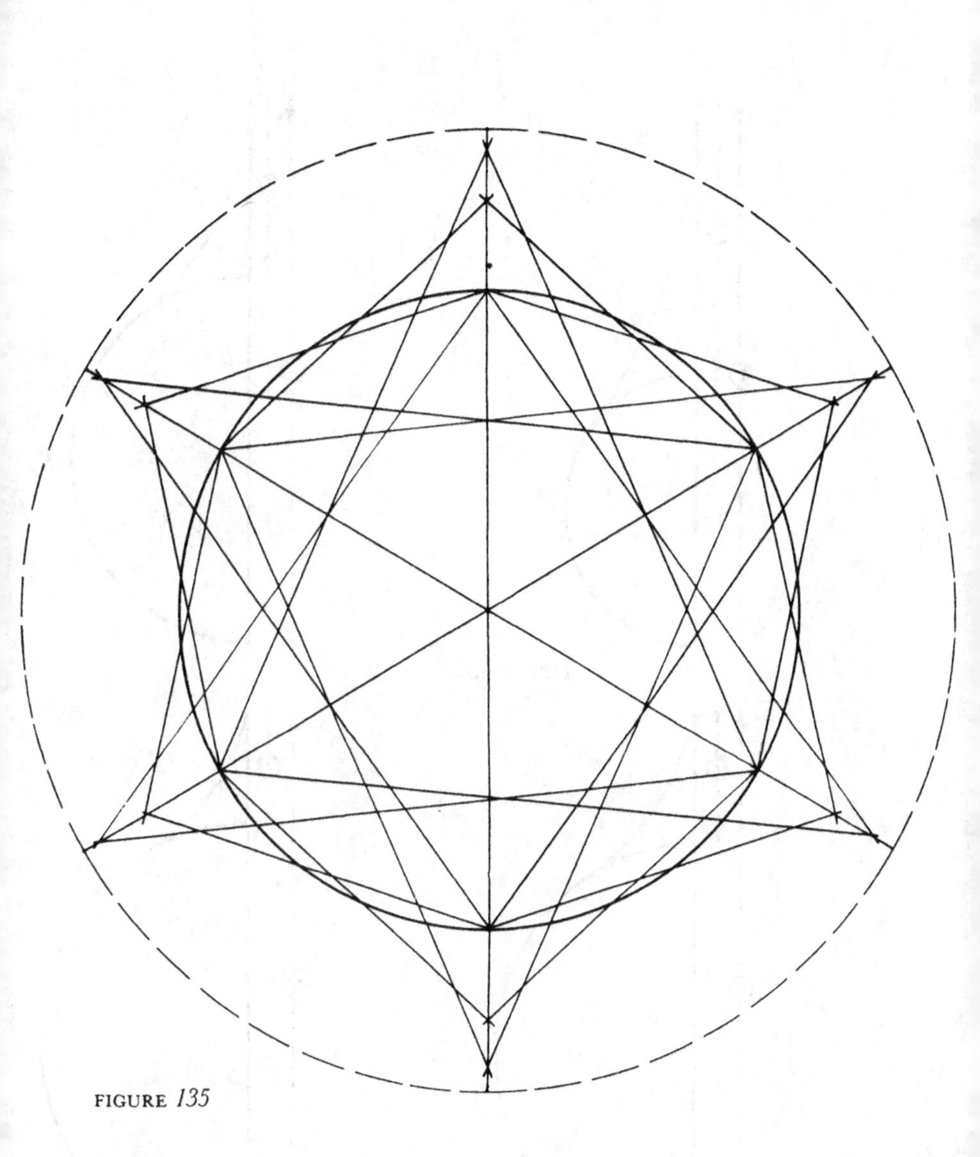

FIGURE *135*

confirmed by letting one end of the chord go 5 times faster than its partner and producing the expected 6-pointed star of Figure 135.

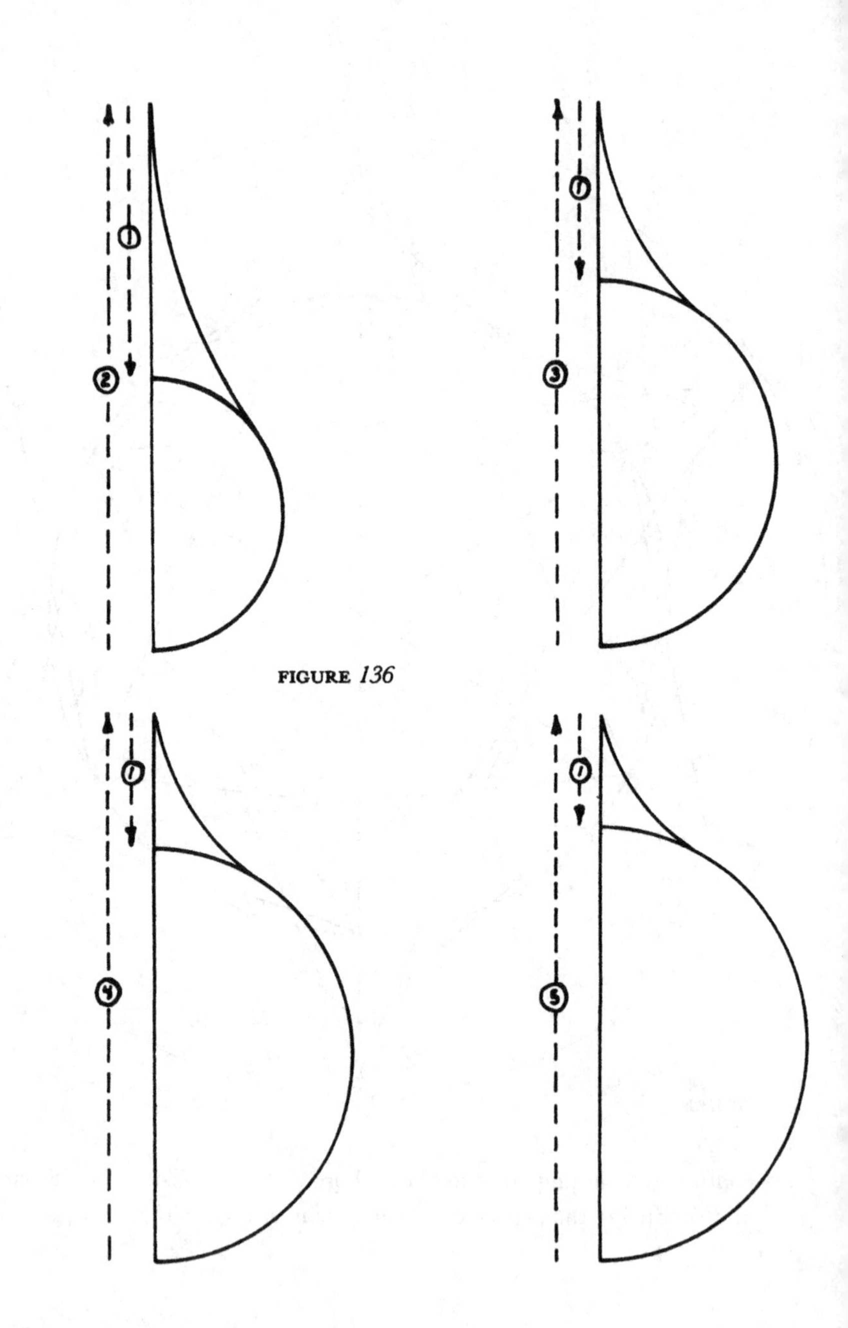

FIGURE 136

A comparison of Figures 132 to 135 reveals that, in relation to the radius of the parent circle, the cusps are getting shorter as they become more numerous. In the case of the deltoid, the distance of the cusp from the circumference seems equal to the diameter. If this is so, does it tie up an any way with the speed-ratio of 2:1 that produced the figure? Measurement shows that the cusp divides the diameter externally in the ratio 2:1. The cusps in the other three drawings divide the diameters externally in the same ratios as those of the speeds of the ends of the traveling chords. Figure 136 makes these relationships clearer.

What has started as the recreational meanderings of a chord in its circle has led to mathematical art and artful mathematical relationships.

BOUNCING BILLIARD BALLS

What is the path of a billiard ball when it is allowed to bounce off the sides of various kinds of pool tables? Along with the usual rectangular pool table, other tables should be considered: regular and irregular polygonals; those with curved sides; *rooms* from whose walls the ball is bounced. Also to be considered is the type of path to be studied: the shortest path to a given point on the table after n bounces; the path that will return the ball to its starting point; paths that yield equal segments of path length between successive bounces; and so on.

Let there be a regulation pool table $ABCD$ (Figure 137a) with two balls, P and Q. If Q is to be hit with P, then P is sent toward Q by hitting P with the cue. This would certainly produce the shortest path from P to Q. Suppose, however, we wish to find the shortest path to Q after one or more ricochets from the sides of the pool table.

It is relatively simple to determine such a path. There is a choice of 4 cushions, or sides, and the 4 possible paths utilizing only one bounce are shown in Figure 137b. The ball rebounds so that the angle of incidence, θ_1, is equal to the angle of reflection, θ_2. This law of reflection is fundamental to the solution; the results of putting "English" on the ball

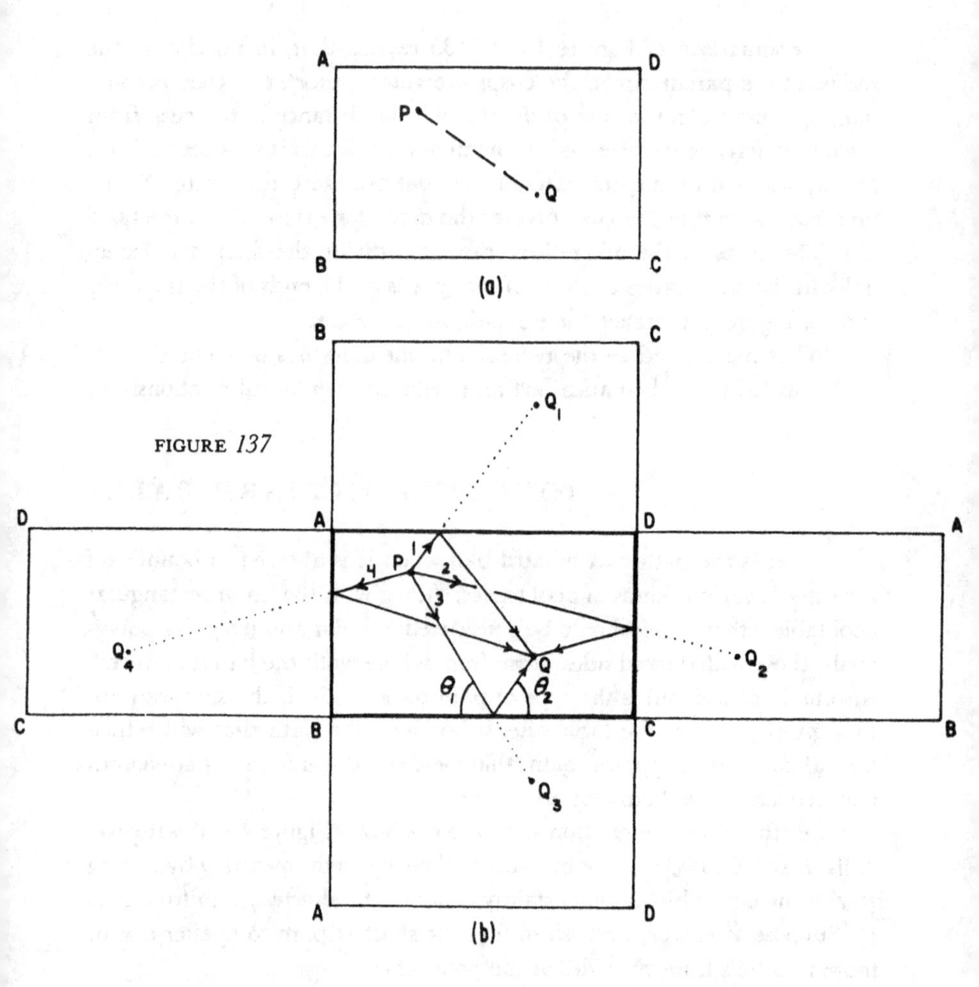

FIGURE *137*

are not considered. Mirror images of the table are constructed at each cushion; a straight line drawn from P to the various mirror images of Q; and a line to the real Q, is drawn from the point of intersection of the straight line with the cushion. That path 1 is the shortest can be confirmed by measurement. However, a simple rule for such cases (one bounce, a rectangular table) is that the shortest path from P to Q is the rebound path from the cushion closest to P or Q. Cushion BC is closest to Q, but cushion AD is even closer to P, so cushion AD is the rebound cushion and path 1 is the shortest of the four paths.

If the ball is to rebound from 2 or more cushions, the situation is a bit more complicated, but Figure 138 shows the double-rebound case with path 1 the shortest. Two bounces require P to pass through or into 2 mirror images of the table to Q, 3 bounces require 3 mirror images, which can be obtained by extending the pattern of rectangles in Figure 138 indefinitely, and so on.

FIGURE *138*

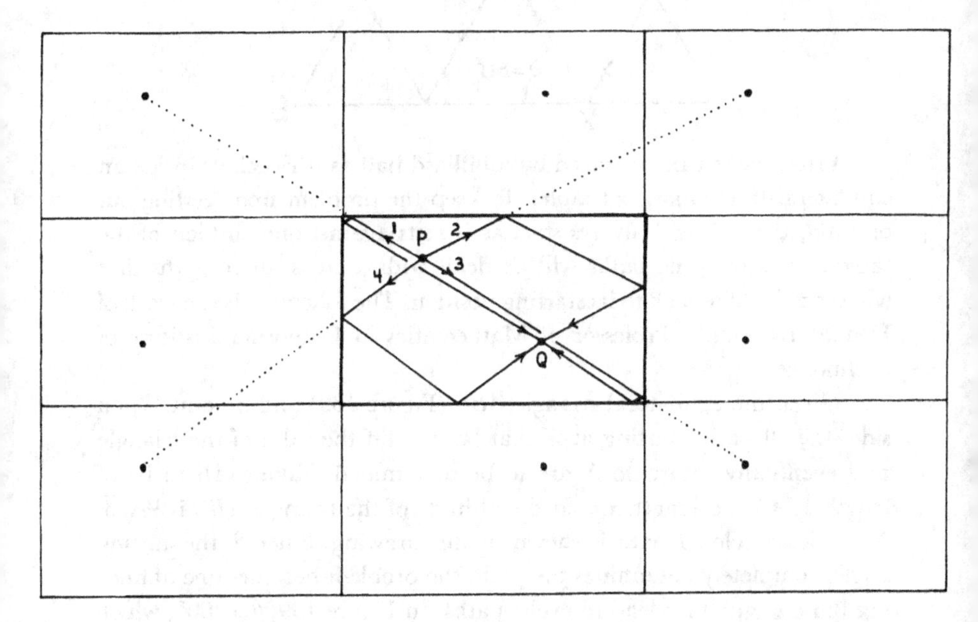

If P is to return to its original position by way of the shortest path, it is bounced perpendicularly off the closest cushion. To return it after 2 or more bounces becomes complicated, for there appears to be a maddening tendency for the ball to enter corners. This problem will be taken up next on a simpler pool table.

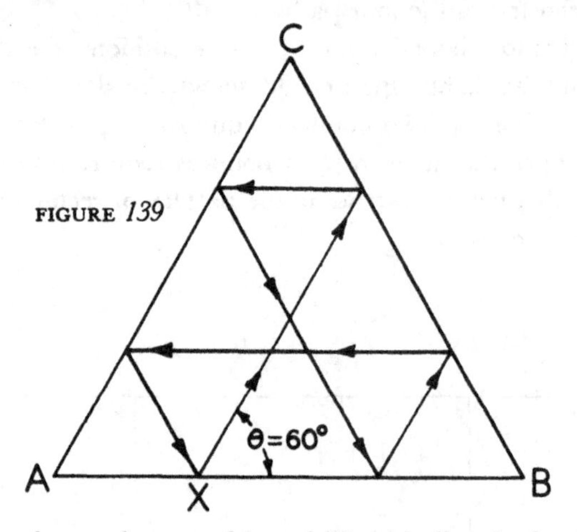

FIGURE *139*

What paths can be traced by a billiard ball as it ricochets inside an equilateral-triangular pool table? To keep the problem from getting out of hand, the ball will always start as it rests against one cushion of the table and only cyclic paths will be dealt with, that is, those paths that will send the ball back to its starting position. The following is the work of Donald E. Knuth, Professor of Mathematics at California Institute of Technology.

Given the equilateral triangle ABC (Figure 139) and a point, X, on side AB, all paths starting at X that bounce off the sides of the triangle and eventually return to X are to be determined. Taking AB to be of length 1, AX has length x, and the altitude of the triangle ABC is $\frac{1}{2}\sqrt{3}$. A particular closed path is shown in this drawing. Since θ, the starting angle, completely determines the path, the problem becomes one of finding those angles that lead to cyclic paths. In Figure 139, $\theta = 60°$, which

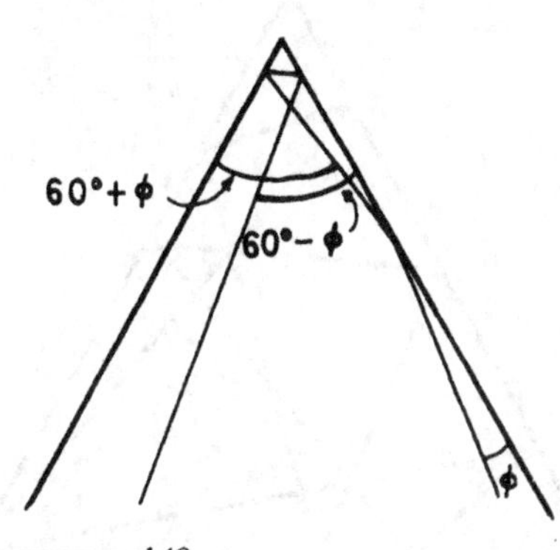

FIGURE *140*

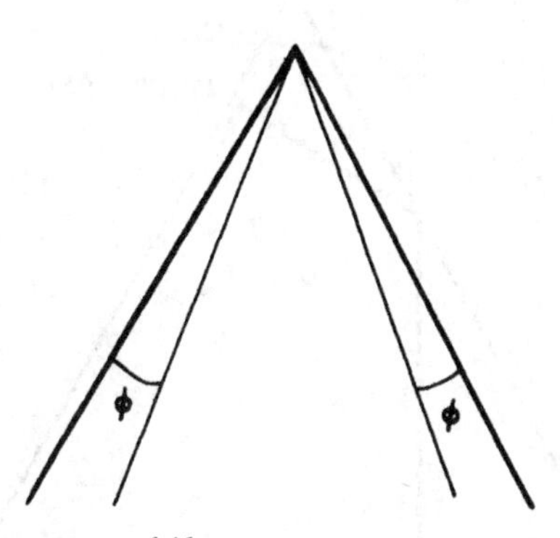

FIGURE *141*

FIGURE *142*

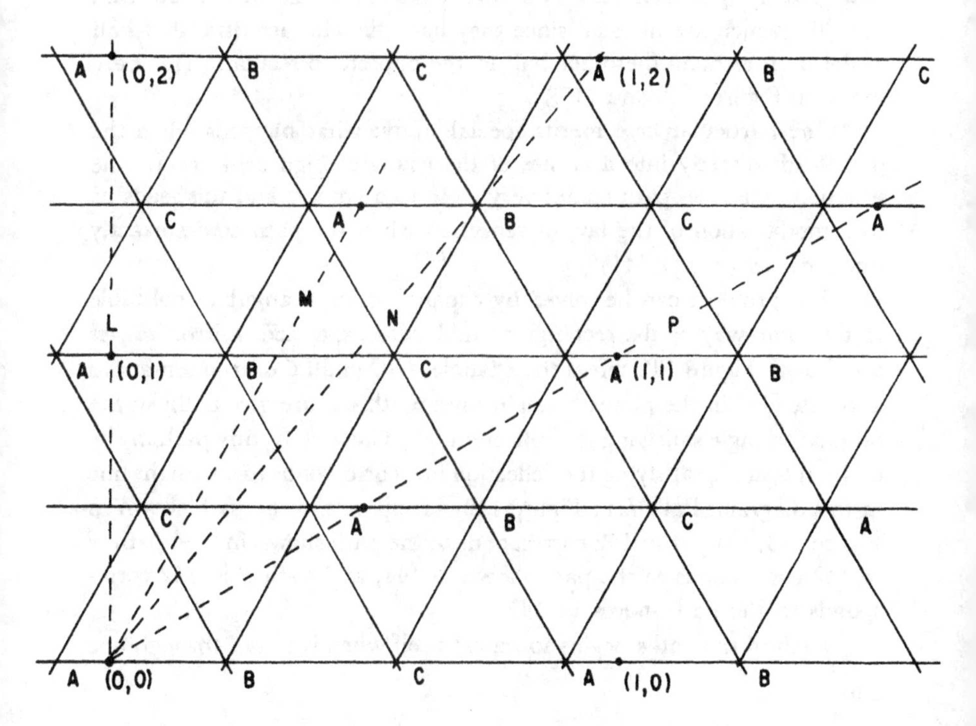

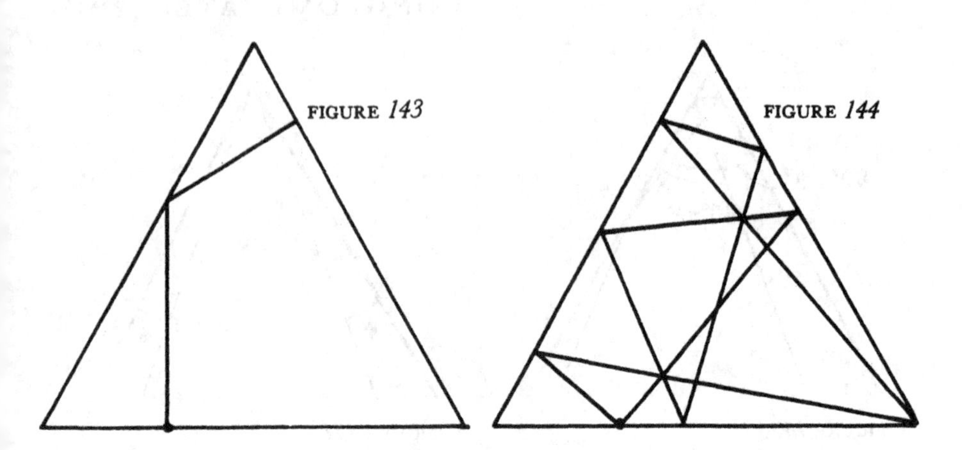

is an interesting special case. Two other cases of interest are $\theta = 30°$ and $\theta = 90°$, which are unusual since they have the characteristic that half of the path retraces the other half in the opposite direction. (These are shown in Figures 143 and 145.)

One particular case merits special study: what happens when the path leads directly into a corner of the triangle? Figure 140 shows the situation when the path comes very close to a corner, and this leads us to a modification of the law of reflection when the path comes exactly into a corner (Figure 141).

The problem can be solved by expanding the triangular pool table in the same way as the rectangular table was expanded: mirror images are drawn. Figure 142 shows the triangle *ABC* in all 6 of its orientations and extended in the plane. Straight lines on this figure give paths in the original triangle satisfying the reflection law. Conversely, any path in the original triangle satisfying the reflection law corresponds to a straight line in this diagram. Path *L* in Figure 142 corresponds to the path shown in Figure 143; path *M* in 142 corresponds to the path shown in 139; path *N* in 142 corresponds to the path shown in 144; and path *P* in 142 corresponds to the path shown in 145.

Each of the paths begins to repeat itself when it passes through one

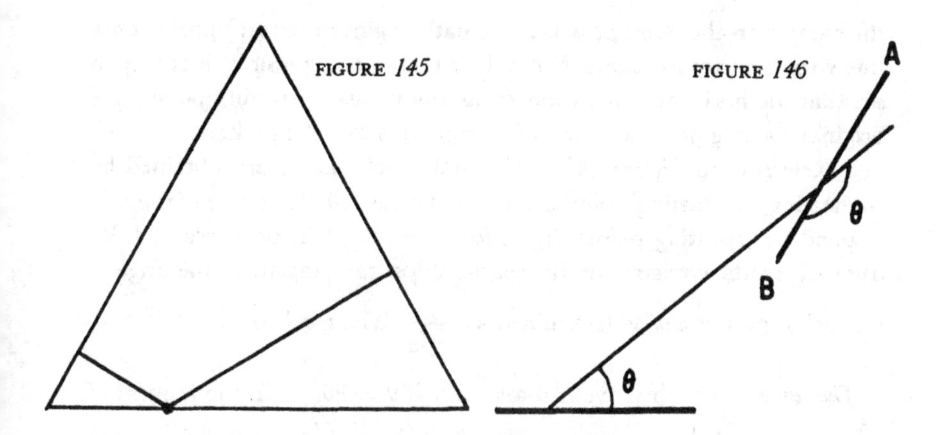

of the copies of the starting position. The angle θ is restricted to values greater than 0° and less than or equal to 90°, otherwise paths that are merely the same ones traced in the opposite direction are found. The point at which the path begins to repeat might conceivably appear on a line with a left to right slope. But this leads to the impossible situation shown in Figure 146, where the corresponding angles are not equal as they should be. A second possibility is that the point at which the path begins to repeat might appear on a line with a right to left slope. But Figure 147 shows that this is possible only when $\theta = 60°$ and $x = ½$ (where x is the distance AX as noted in Figure 139). We are now left with

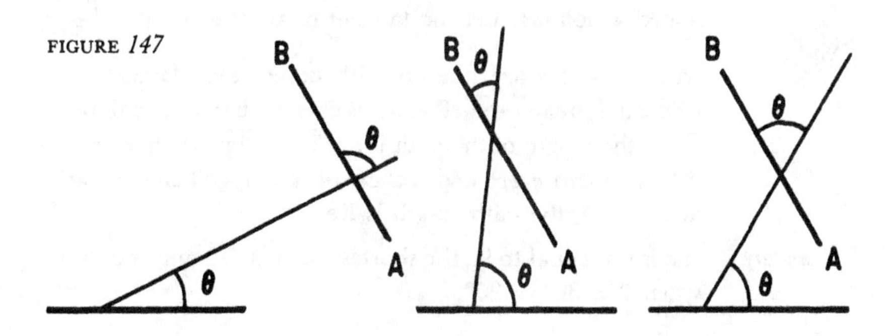

the case when the point at which the path begins to repeat appears on a line with a horizontal slope. Not only is such a case possible, it is easy to see that the horizontal line connecting any two such starting points (the original starting point and one of its copies) is a cyclic path.

Referring to Figure 142, all possible cyclic paths are obtained by connecting the starting point, arbitrarily labeled (0, 0), to one of the corresponding repeating points, (i, j), for example (0, 2) on a side AB. By using elementary trigonometric relationships, the tangent of the angle θ for such a path is easily determined as $\frac{j}{i\sqrt{3}}$. The result is:

Theorem 1: A path is cyclic if and only if $\theta = 90°$, or if the tangent of the angle θ is equal to $\frac{r}{\sqrt{3}}$, where r is a non-0 and rational number. A *rational number* is a number that can be expressed as the quotient of one whole number divided by another whole number. Numbers that cannot be so expressed, for example, $\sqrt{2}$, are called irrational numbers.

Let $r = \frac{p}{q}$ be the rational number expressed as a fraction reduced to its lowest terms. Then, if both p and q are odd, the path passes first through the point $\left(\frac{q}{2}, \frac{p}{2}\right)$, otherwise it passes first through (q, p). This leads to:

Theorem 2: The length of the path traveled in each cycle may be determined as follows: Let the tangent of angle θ equal $\frac{p}{(q\sqrt{3})}$ where p and q are integers with no common factor, and where p is greater than 0 and q is greater than or equal to 0. Then the length of the path is $k\sqrt{3p^2 + 9q^2}$ where $k = ½$, if both p and q are odd; otherwise $k = 1$. When $\theta = 60°$ and $x = ½$, the path length is 1.5.

Corollary: If x is not equal to ½, the shortest path is $\sqrt{3}$ and it occurs when $\theta = 30°$ or $90°$.

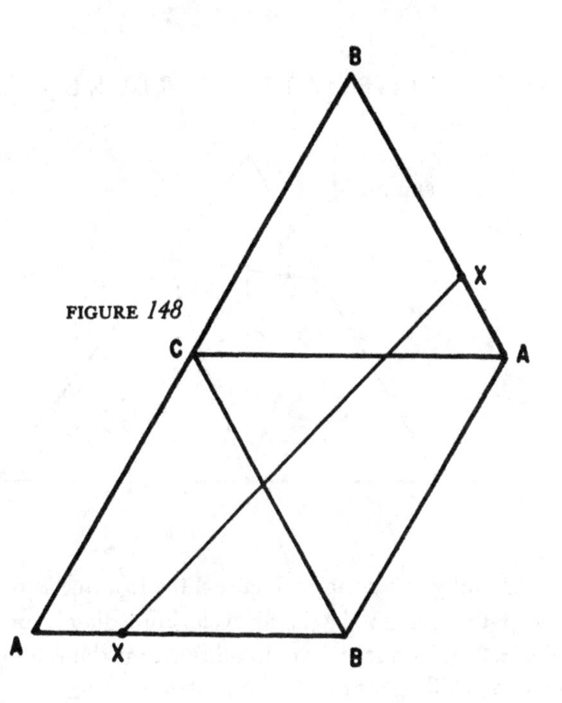

FIGURE *148*

Next comes the number of bounces in each path. To account for the degenerate cases when the path leads directly into a corner, that event will be counted as 3 bounces, as suggested by Figure 140. The return of the ball to the starting point is counted as a bounce.

Theorem 3: The number of bounces occurring in each cycle may be determined as follows: Let p, q, k be as in Theorem 2, then the number of bounces is:

$k(2p + 6q)$ (for p greater than or equal to $3q$)

or $k(4p)$ (for p less than or equal to $3q$)

except when $\theta = 60°$ and $x = ½$, when the number of bounces is 3.

Corollary: If x is not equal to ½, the least number of bounces per cycle is 4 and this occurs when $\theta = 30°$ and $90°$. The number of bounces is always even.

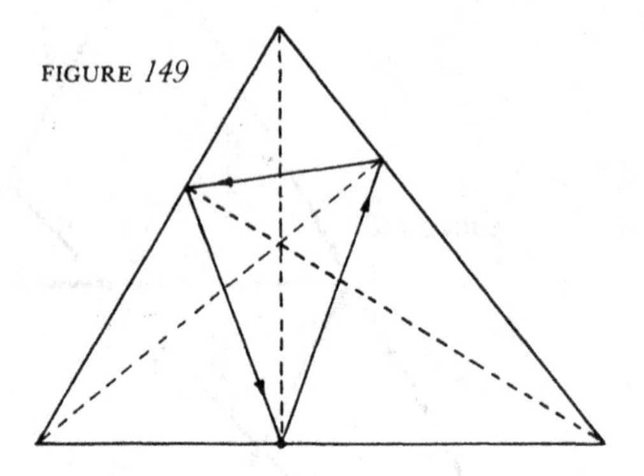
FIGURE 149

Finally, referring to Figure 148: If a line is drawn between the X's, the path does not begin to cycle immediately unless $x = ½$, as shown above. Now, under what circumstance is the path cyclical if the figure is extended? Tangent of θ is computed as

$$\frac{1}{\sqrt{3}}\left(\frac{1+x}{1-x}\right)$$

and $\dfrac{(1+x)}{(1-x)} = r$ must be rational. This implies that $x = \dfrac{(r-1)}{(r+1)}$ and finally results in:

Theorem 4: The line in Figure 148 leads to a cyclic path if and only if x is rational.

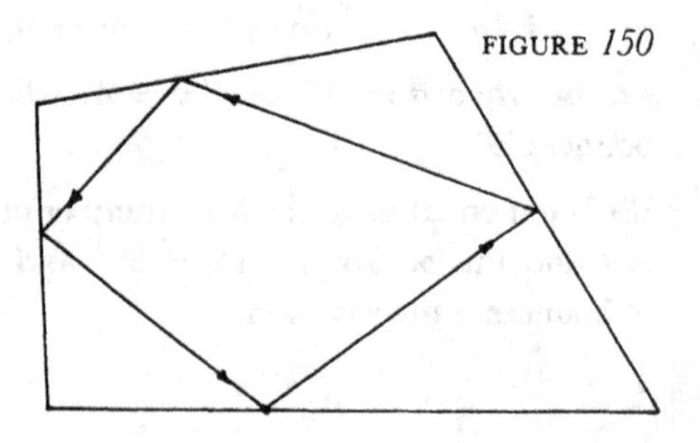
FIGURE 150

The cyclic path of a ball within an acute-scalene triangular pool table is easily found, and it turns out to be the pedal triangle of Figure 149. A pedal triangle is formed by the feet of the altitudes of a triangle.

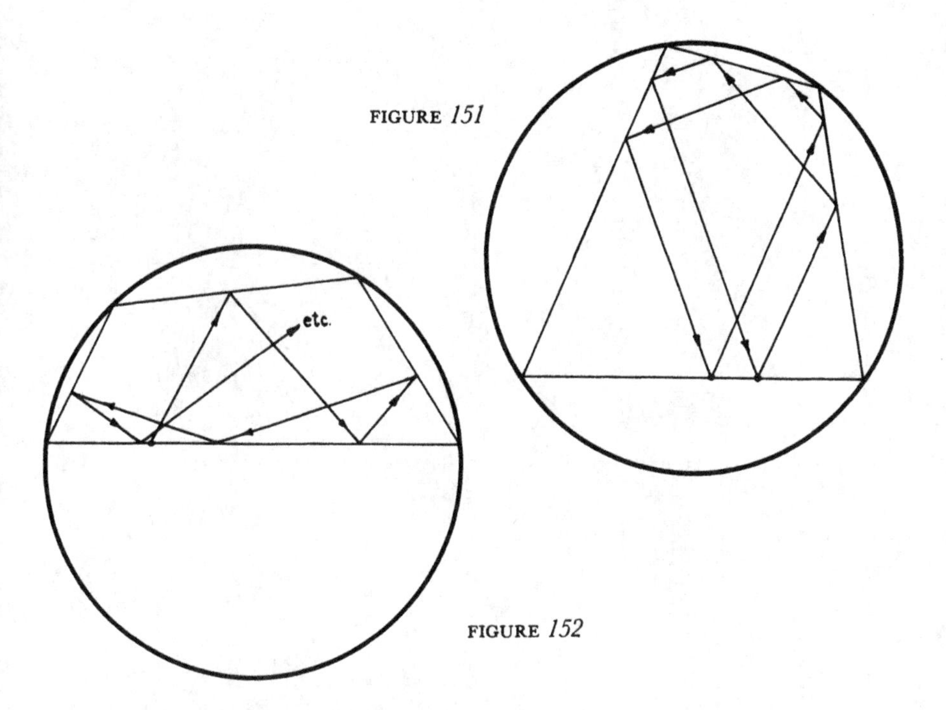

FIGURE *151*

FIGURE *152*

Can a cyclic path be found in a quadrilateral pool table in which no 2 sides are parallel? Figure 150 shows such a cyclic path for a particular case, but is it true for *all* such quadrilaterals?

Interestingly enough, such paths are possible if the quadrilateral is inscribable in a circle *and* if the center of the circle lies within the quadrilateral. Figure 151 shows a number of cyclic paths in another quadrilateral of the type specified, and Figure 152 shows a quadrilateral for which no cyclic path is possible.

GLOSSARY

Aliquot Divisor An integer, smaller than a given number, *N*, such that when it is divided into *N*, another integer is produced.

Alphametic A mathematical problem in which the digits are replaced by letters or words that form sensible words or phrases.

Amicable Number One of a pair of integers so related that the sum of the aliquot divisors of one of the integers is equal to the other integer.

Antimagic Square An array of integers, usually consecutive starting with unity, such that the sums of the integers in each row, column, or long diagonal form a series of consecutive integers.

Automorphic Number An integer whose square ends with the given integer.

Combination Any given selection of one or more items in a set of objects. Some of the various combinations of the 4 letters, *a*, *b*, *c*, and *d* are *abc*, *cad*, *abd*, *a*, *cd*, and *abcd*.

Composite Number An integer that is not a prime number.

Cryptarithm A mathematical puzzle in which the digits are replaced by any other symbols.

Digit In the base-10 system, any single one of the individual symbols used to express quantities.

Digital Invariant An integer equal to the sum of some *n*th power of its digits, or group of digits.

Directrix See parabola.

Ellipse A curve, like a slightly flattened circle, that is generated by a point moving so that the sum of the distances from the moving point to two fixed points, called the *foci*, is constant.

Elliptipool Pool played on an elliptically shaped pool table.

Factor Any one of two or more numbers that yield a product when multiplied together. Factors of 12 are 3 and 4, or 2 and 6, or 2, 3, and 4.

Factorial of a Positive Integer The product of all the positive integers from unity to the given integer is termed a factorial. The symbol for factorial *n* is *n*!.

Flexagons Plane structures made of folded paper than can be turned inside out several times to shown different faces.

Flexahedrons Three-dimensional structures made up of connected solids that can be rotated or flexed in various ways.

Focus See ellipse, parabola.

Integer A whole number, usually positive.

Magic Constant See magic square.

Magic Square An array of numbers such that the sum, called the *magic constant,* of the numbers in each row, column, and long diagonal is the same.

Multiperfect Number An integer such that the sum of its aliquot divisors, plus the given integer, is equal to a whole multiple of the integer itself.

Narcissistic Number An integer that is representable, in some way, by mathematically manipulating the digits of the integer.

Number A mathematical symbol denoting a quantity. Numbers may be whole, fractional, decimal, imaginary, complex, and so on.

Origami The Japanese art of folding paper to form specific objects.

Parabola A curve generated by a point that moves so that its distances from a fixed point, called the *focus,* and a straight line, called the *directrix,* are equal.

Perfect Number An integer such that the sum of its aliquot divisors is equal to the integer.

Permutation An ordered arrangement or sequence of several things.

Polygon Any area enclosed by a series of connected straight line segments.

Prime Number An integer having only unity as an aliquot divisor.

Quadrille A pattern of dominoes in which like numbers of dots are arranged in groups of four.

Symmetry Similarity of form or arrangement on either side of a dividing line.

Talisman Squares An array of integers such that the difference between any given integer and its neighbors are greater than some given constant.

BIBLIOGRAPHY

Many books and magazines contain excellent chapters, sections, or articles in the field of recreational mathematics; some books cover the subject thoroughly, and I have listed a few of these. The only recent magazines in English solely devoted to the field are *Recreational Mathematics Magazine,* which ceased publication in January 1964 and the *Journal of Recreational Mathematics. Sphinx,* published in French from 1931 to 1939, is available in so few libraries that reference to it would be rather difficult.

The criterion for this Bibliography has been the books I found myself using most often in connection with the topics dealt with in this book.

ANDREWS, W. S. *Magic Squares and Cubes.* New York: Dover Publications, Inc., 1960.

This classic reference work includes hundreds of examples and dozens of methods of constructions of magic squares.

BALL, W. W. ROUSE. *Mathematical Recreations and Essays.* Revised by H. S. M. COXETER. Toronto: University of Toronto Press, 1974.

This basic work was first published in 1892 by the Macmillan Company and has undergone twelve editions or revisions.

BEILER, ALBERT H. *Recreations in the Theory of Numbers—The Queen of Mathematics Entertains.* New York: Dover Publications, Inc., 1964.

A veritable gold mine of number recreations.

BERNHART, ARTHUR. "Curves of Pursuit," *Scripta Mathematica,* XX (September-December, 1954), 125–141.

———. "Curves of Pursuit—II," *ibid.,* XXIII (1957), 49–65.

———. "Polygons of Pursuit," *ibid.,* XXIV (Spring, 1959), 23–50.

———. "Curves of General Pursuit," *ibid.,* III (September, 1959), 189–206.

These four articles discuss virtually every type of pursuit situation and include a great number of historical references.

BROOKE, MAXEY. *Fun For the Money.* New York: Charles Scribner's Sons, 1963.

A collection of puzzles and games using coins.

———. *150 Puzzles in Crypt-Arithmetic.* New York: Dover Publications, Inc., 1963.

The only book entirely devoted to alphametics and cryptarithms, giving general hints to aid in solving the puzzles.

DUDENEY, H. E. *Amusements in Mathematics.* New York: Dover Publications, Inc., 1958.

———. *The Canterbury Puzzles.* New York: Dover Publications, Inc., 1958.

These two books contain just a part of the output of ingenious puzzles by Dudeney.

GARDNER, MARTIN. *Mathematics, Magic and Mystery*. New York: Dover Publications, Inc., 1956.

Card tricks, tricks with everyday objects, topological effects, geometrical vanishes, and other magic based on mathematical principles.

———. *The Scientific American Book of Mathematical Puzzles & Diversions*. New York: Simon and Schuster, Inc., 1959.

———. *The 2nd Scientific American Book of Mathematical Puzzles & Diversions*. New York: Simon and Schuster, Inc., 1961.

———. *The Incredible Dr. Matrix*, New York: Charles Scribner's Sons, 1976.

———. *Mathematical Magic Show*, New York: Alfred A. Knopf, 1977.

These excellent volumes contain the cream of Gardner's monthly columns in *Scientific American* with new material supplied by his readers.

HUNTER, J. A. H., and MADACHY, JOSEPH S. *Mathematical Diversions*. New York: Dover Publications, Inc., 1975.

———. *Figurets: More Fun With Figures*. Toronto: Oxford University Press, 1958.

———. *Mathematical Brain-Teasers*. New York: Dover Publications, Inc., 1976.

These three books are but a small sample of Hunter's prolific outpouring.

———. "Two Very Special Numbers," *The Fibonacci Quarterly*, II (October, 1964), 230.

The two 100-digit automorphic numbers discovered by R. A. Fairbairn.

HUNTER, J. A. H., and MADACHY, JOSEPH S. *Mathematical Diversions*. Princeton: D. Van Nostrand Co., Inc., 1963.

This book ranges over a wide field of recreational mathematics and offers a collection of puzzles and alphametics.

KASNER, EDWARD, and NEWMAN, JAMES R. *Mathematics and the Imagination*. New York: Simon and Schuster, Inc., 1940.

One of the classic standards about mathematics and its more light-hearted aspects.

KRAITCHIK, MAURICE. *Mathematical Recreations*. New York: Dover Publications, Inc., 1953.

Kraitchik's source of material stems from his eight years as editor of *Sphinx*.

LINDGREN, HARRY. *Geometric Dissections*. Princeton: D. Van Nostrand Co., Inc., 1964; revised and enlarged by GREG FREDERICKSON, New York: Dover Publications, Inc., 1972.

This book is the only one solely devoted to its subject, containing most of the known nontrivial dissections and some systematic techniques for finding new dissections. Puzzles are posed for the reader who wants to indulge in the pastime.

LOYD, SAM. *Mathematical Puzzles of Sam Loyd*, ed. MARTIN GARDNER. New York: Dover Publications, Inc., 1959.

———. *Mathematical Puzzles of Sam Loyd—Volume Two*, ed. MARTIN GARDNER. New York: Dover Publications, Inc. 1960.

Loyd, like Dudeney, was a composer of ingenious puzzles.

NORTHROP, EUGENE P. *Riddles in Mathematics—A Book of Paradoxes.* Princeton: D. Van Nostrand Co., Inc., 1944.

How to prove that all triangles are isosceles, that there are as many points on a line segment as there are on a whole line, and others of this type of mathematical riddle. This book introduced me to mathematical recreations.

OAKLEY, C. O., and WISNER, R. J. "Flexagons," *American Mathematical Monthly,* LXIV (March, 1957), 143–154.

An extended mathematical treatment of flexagons.

PHILLIPS, HUBERT [CALIBAN]. *My Best Puzzles in Mathematics.* New York: Dover Publications, Inc., 1961.

A collection of the popular mathematical puzzles by this famous English puzzlist.

RUMNEY, MAX. "Digital Invariants," *Recreational Mathematics Magazine,* No. 12 (December 1962), 6–8.

SHAFF, WILLIAM L. *A Bibliography of Recreational Mathematics,* Volume 1 (1970), Volume 2 (1970), Volume 3 (1973). Washington, D.C.: National Council of Teachers of Mathematics.

This is *the* bibliography in which one may find a multitude of references to any topic in recreational mathematics. Over 7,000 references are grouped into thirteen major classifications.

SPRAGUE, ROLAND. *Recreation in Mathematics—Some Novel Problems.* Translated by T. H. O'BEIRNE. London: Blackie and Son Limited, 1963.

STEINHAUS, HUGO. *Mathematical Snapshots.* New York: Oxford University Press, 1969.

Graphic and pictorial treatment of many important topics in mathematics.

"Problem 86," *Mathematics Magazine,* XXIV (January-February, 1951), 166.

This is probably the first appearance of the idea of antimagic squares, called heterosquares in this reference.

INDEX

ABOUT THE AUTHOR

Joseph S. Madachy's interest in recreational mathematics started during his high school years. Although trained in chemistry, in which he held B.S. and M.S. degrees from Western Reserve University, he taught algebra and physical science, as well as chemistry, at Cathedral Latin High School in Ohio and was an Instructor in Trigonometry at Western Reserve University.

Madachy was a research chemist and a technical editor for Monsanto Research Corporation until his retirement in 1988. He was the editor of the *Journal of Recreational Mathematics* for nearly 30 years and then served as editor emeritus. He made numerous contributions to the field of recreational mathematics during his lifetime.

Madachy passed away in 2014, at the age of 87, near Kettering, a suburb of Dayton, Ohio, where Madachy and his wife, Juliana, who married in 1957, had raised a family of six children.